Marian Mikołajczuk

Colour profiles

Artur Juszczak, Marek Radomski

Yakovlev Yak-23

The first Yakovlev jet fighters.

STRATUS

Published in Poland in 2008
by STRATUS s.c.
P.O. Box 123,
27-600 Sandomierz 1, Poland
e-mail: office@mmpbooks.biz
for
MMP,
36 Ver Road, Redbourn,
AL3 7PE, UK.
e-mail: rogerw@mmpbooks.biz

http://www.mmpbooks.biz

ISBN
978-83-89450-54-8

Editor in chief
Roger Wallsgrove

Editorial Team
Bartłomiej Belcarz
Artur Juszczak
James Kightly
Robert Pęczkowski
Michał Mietelski

Translation
Wojtek Matusiak

Colour Drawings
Marek Radomski
Artur Juszczak

Scale plans
Dariusz Karnas

DTP
Robert Pęczkowski
Artur Juszczak

Printed by:
Drukarnia Diecezjalna,
ul. Żeromskiego 4,
27-600 Sandomierz
tel. +48 (15) 832 31 92;
fax +48 (15) 832 77 87
www.wds.pl marketing@wds.pl

PRINTED IN POLAND

Table of contents

Yak-15 design and development 4
Birth of the Yak-17 20
Yak-17 successors 30
Yak-23 development 32
Yak-17 and Yak-23 exports 40
  Poland 40
  Czechoslovakia 50
  Bulgaria 52
  Romania 56
  People's Republic of China 60
  Hungary 60
Yak-23 structure 61
  Specification 63
  Short evaluation of the Yak-23 64
Bibliography: 68
Detail photos 69
  Yak-15 69
  Yak-17 83
  Yak-17UTI 90
  Yak-23 94
Colour profiles 119

Acknowledgements

Author and publisher would like to thanks following people:
Dmitri Sribnyi - editor of www.airforce.ru, Michal Ovcacik - 4+ Publishing Co., Mirosław Czaplicki, Dénes Bernád, Stefan Boshniakov, Andrzej Gołąbek, Krzysztof Radwan - MLP, Andrzej Morgała, R. Karpiński, Andrzej Tatarek, Robert Gretzyngier, Piotr Słoma, Stanisław Kopacz, Franciszek Dziekan, Wacław Hołyś, Anna Garbacz.

Get in the picture!
Do you have photographs of historical aircraft, airfields in action, or original and unusual stories to tell?
MMP would like to hear from you! We welcome previously unpublished material that will help to make MMP books the best of their kind. We will return original photos to you and provide full credit for your images. Contact us before sending us any valuable material: rogerw@mmpbooks.biz

Title page: *Yak-23 of the Polish Air Force, Tomaszów Mazowiecki airfield.*
F. Dziekan.

Early Yakovlev jets development

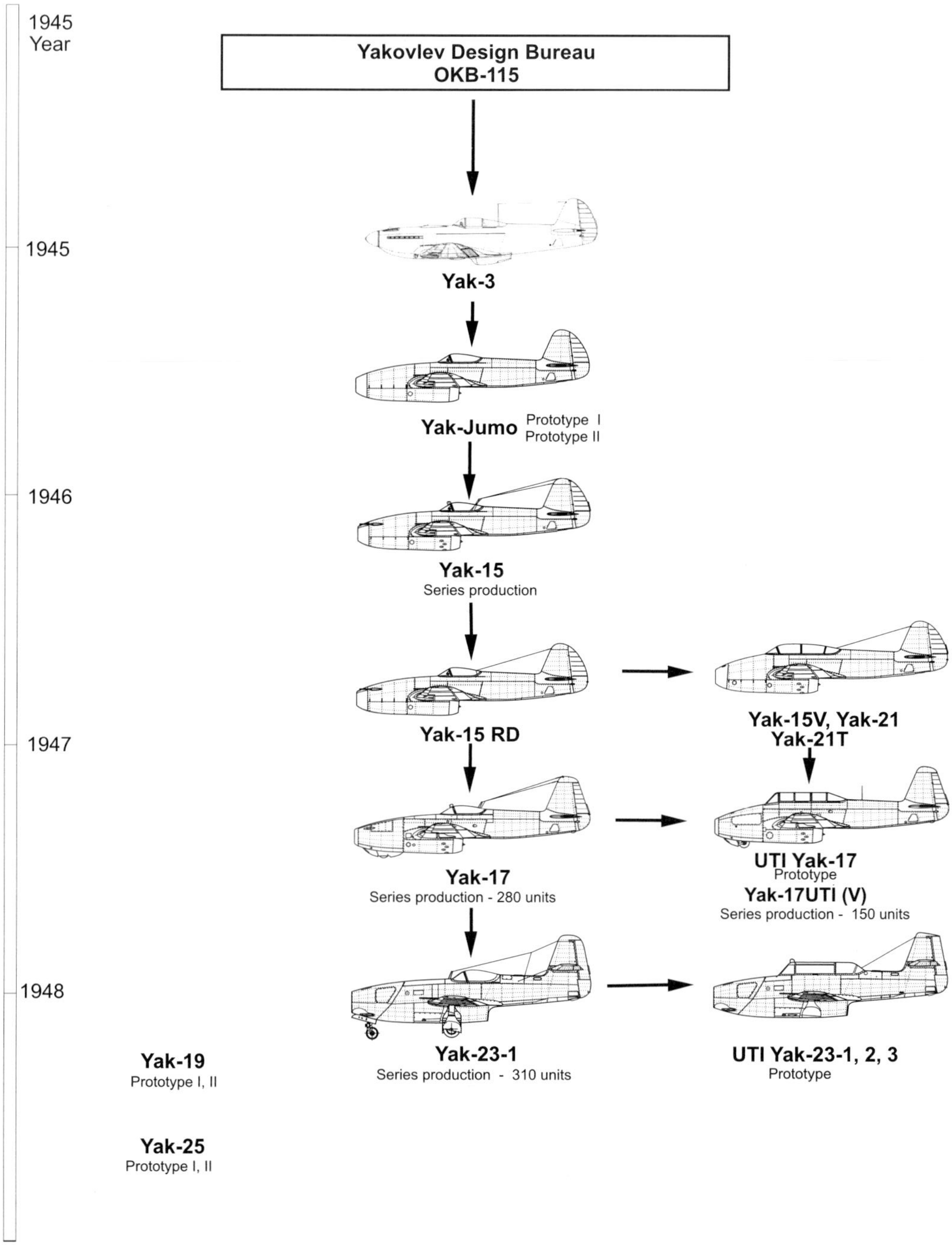

Yak-15 design and development

Over the centuries every armed conflict has initiated and enhanced rapid developments in engineering. The two world wars of the 20th century were no exception. World War Two was a particular period of technological progress. New technologies were developed on both sides of the conflict, altering the nature of the war and accelerating its end.

It was in Germany that the first turbojet (Messerschmitt Me 262, Arado Ar 234) and rocket powered (Messerschmitt Me 163) aircraft were introduced into front line service. New, more advanced aircraft were also under development. The end of hostilities in Europe provided the winners with many war prizes in the form of new armament types, including aircraft. This was particularly important for the USSR, which had been mostly isolated from the Western world since its establishment.

Work on jet aircraft had started in the USSR in the 1930s, but with little progress. It was the end of the war and the influx of German experience and technologies that led to increased interest in the subject. On 9 April 1945 the National Defence Committee of the USSR issued a directive to begin work to develop a jet-powered aircraft. This was planned to be a fighter aircraft powered by a captured Junkers Jumo 004B1 or BMW 003 jet engine. At the time the decisive parameter of the fighter was its maximum speed, and this was set at 800 km/h (500 mph). In parallel with work on the aircraft, the Jumo 004B1 was used as a pattern to develop a Soviet power plant designated RD-10 (*Reaktivnyi Dvigatel* - Jet Engine). By the orders of the Ministry of Aircraft Production on 28 April 1945, design work on the power plant was entrusted to OKB-117 (*Opitnoye Konstruktorskoye Byuro* – Experimental Design Bureau) headed by the chief designer Vladimir Y. Klimov. Production of RD-10 engines was going to be undertaken by No. 26 Plant at Ufa. Work on the engine was led unofficially by Nikolai D. Kuznetsov, the first deputy to the head. Three months later the

Mikoyan I-250 prototype. The aircraft used an additional VRDK engine. ***Stratus coll.***

same ministry prepared another order, issued on 13 June, that called for development of another jet engine designated RD-20, based on the German BMW 003 engine, and putting it in series production. The work was entrusted to the OKB of No. 16 Plant at Kazan, headed by the chief designer Sergey D. Kolosov. It took until the end of 1945 to prepare the documentation and at the beginning of 1946 production of the first batches of these engines was undertaken. The first power plants manufactured had rather limited characteristics, due to the novel technology, and their time between inspections was short and their reliability rather low.

Some work on the jet- and rocket-powered aircraft had been carried in the USSR before and during WW2. Leading fighter aircraft designers attempted to fit their designs with additional rocket boosters. Artem I. Mikoyan and Mikhail I. Gurevich developed an interceptor designated I-250, fitted with an additional VRDK engine (*Vozdushnoreaktivniy Dvigatel s Kompresorom* - air jet engine with compressor). A similar powerplant was adopted in the designs of Pavel O. Sukhoi (Su-5 experimental fighter), and the MiG design saw limited production and service post-war as the MiG-13. Rocket and ramjet boosters were tested by other designers, such as Semyon A. Lavochkin (La-7R, 120R and 126 fighters) and Alexandr S. Yakovlev (Yak-3RD with the RD-1 engine and Yak-7PVRD with two DM-4 ramjets). Work on these failed to produce satisfactory results for practical operational aircraft. The general concept was flawed and it was the captured German technology that encouraged leading designers of individual bureaux to undertake bolder solutions.

Based on the directive issued in April, work on development of a jet-powered fighter aircraft was undertaken by several design offices:

- OKB-155 headed by A. I. Mikoyan and M.I. Gurevich;
- OKB-301 under S. A. Lavochkin;
- OKB-21 of Semyon M. Alekseyev;
- OKB-51 of P. O. Sukhoi;
- OKB-115 under A. S. Yakovlev.

Rapid progress in research and development work, and construction of jet-powered aircraft in Allied countries (Britain and the USA), made Soviet leaders realise the technological gap between the USSR and the West. Therefore, it was clear from the outset that the leaders of the USSR expected quick results in this prestigious category, so the design offices had to adopt simple solutions.

Yak-3RD experimental mixed power fighter. The rocket exhaust beneath the rudder is covered by a pointed fairing.
Stratus coll.

Su-9 fighter (The izdeliye K, 1945).
Stratus coll.

The path chosen by P. O. Sukhoi was to follow the German Me 262 design. He prepared a project for a twin-engine aircraft known as Product K (*Izdelye K*), later called the Su-9. A different approach was adopted by OKB-155, who prepared a project with the engine located in the forward fuselage, with the exhaust nozzle located under the centre section. At the time this was a rather innovative solution. The OKB-155 project was designated I-270 or Product 270, and later the MiG-9. A similar concept was followed by S. A. Lavochkin, who designed the Product 150 aircraft, which became known as the La-15 following numerous modifications. The OKB-21 design closely resembled the British Gloster Meteor in layout, with twin wing-mounted engines.

A quite simple and original solution was used in the project of A. S. Yakovlev. He was one of the first chief designers able to inspect the wealth of captured German documents on design of rocket and turbojet-powered aircraft. German designers preferred lightweight fighters with a stepped fuselage design. In this layout the power plant was usually located in the forward fuselage. The new design was based on the excellent Yak-3 piston-engined fighter, the same design as used to test the RD-1 rocket engine. The project used the Yak-3's central and rear fuselage, wings, tail surfaces and the undercarriage, along with most systems and equipment. The power plant was located in the forward fuselage, tilted at 4° 30' to the aircraft's axis. Almost the entire fuel system was redesigned, consisting of four tanks (two wing and two fuselage-mounted), with a total capacity of 680 litres. The cockpit canopy was reshaped, in line with the new fuselage outline. Cockpit equipment was largely unaltered, except for the engine controls. Armament was going to consist of two 23 mm NS-23K cannon with 60 rounds per gun, located in the fuselage above the engine. The project was known as the Yak-Jumo (some documents used the designation of Yak-3 Jumo). The 900 kG Jumo 004 engine was going to be used. The air intake featured a fixed cone. Yevgeniy G. Adler was appointed as the leading designer. Considering the simple design, proven airframe and experience from previous experimental designs, Yakovlev realised he would be one of the first, with a clear edge over his competitors.

Work on the project was completed quickly and construction of the first prototype commenced. According to the time schedule the first prototype was going to be ready in the autumn of 1945, half a year after commencement of design work. Since the project was given priority, the prototype was ready for ground runs by early October. The first example did not have a complete set of equipment and armament. The first tests of the power plant revealed numerous shortcomings and faults in the design. During ground runs of the engine, exhaust gases caused excessive heating of the lower rear fuselage. To cure this it was decided to fit a steel screen to protect the structure from overheating. The aircraft was returned to the plant to remedy the shortcomings and faults. Meanwhile, on 8 December the second prototype was completed, and this included all the improvements, as well as complete equipment, armament and an enlarged tailplane. During the winter it was subject to modifications, as it was earmarked for flying testing. In the second quarter of 1946 the testing commenced. On 9 April the aircraft started ground trials, which made it ready for the maiden flight.

At this time the authorities significantly increased the requirements for the new jet-powered aircraft. The decision resulted from intelligence reports about the progress in aircraft technology in Britain and the USA. According to order no. 162 of the Ministry of Aircraft Production of 27.03.1946, the new aircraft were going to have the following specifications:

- maximum speed near the ground - 770 km/h (477 mph);
- maximum speed at an altitude of 5,000 m (16,250 ft) – 850 km/h (527 mph);
- range, at a speed of Ma 0.9 - 500 km (312 miles);
- endurance, at a speed of Ma 0.9 – 40 minutes;
- maximum endurance – 60 minutes;
- time of climb to 5,000 m (16,250 ft) – 4.5 minutes;
- service ceiling – 14,500 m (47,125 ft);
- armament – 2 x 23 mm cannon.

After the second prototype was prepared for flying trials, on 13 December 1945 test pilot Mikhail I. Ivanov performed the first taxiing and acceleration tests, following which the aircraft was sent to *TsAGI* for testing in the T-101 wind tunnel. During these tests the Yak-Jumo was displayed to representatives

Yak-Jumo photographed at No 115 Factory. Moscow, December 1945.
Yakovlev OKB photo via author.

Yak-Jumo.
1/72 scale.

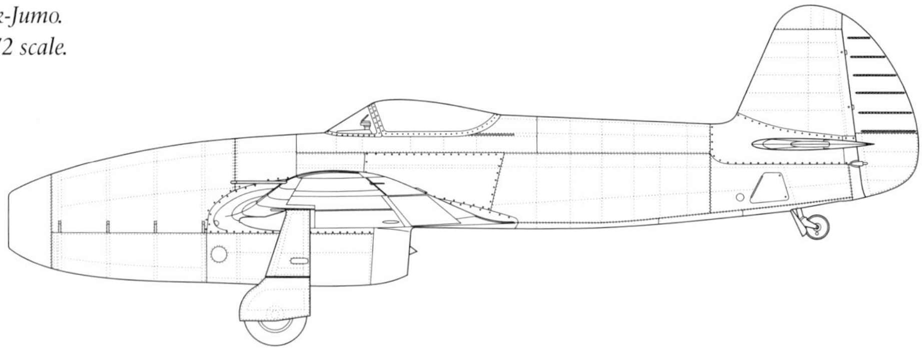

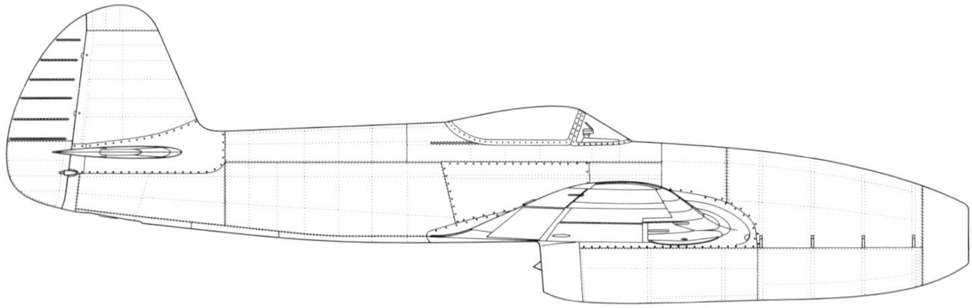

of the Soviet Air Force and designers of principal OKBs. Testing at *TsAGI* continued until February 1946. Upon their completion the aircraft was going to resume ground trials with its power plant, as well as taxiing and acceleration tests. Meanwhile a copy of the Jumo 004 built entirely in the USSR had been delivered, the new engine being designated the RD-10. Its dimensions were the same as those of the original, so it could be fitted in the second prototype with no problem. During these trials problems occurred with the metal tailwheel: during acceleration it was very noisy and emitted sparks. The aircraft was

3/4 front view of the Yak-Jumo, December 1945. ***Yakovlev OKB photo via author.***

returned to the makers to cure the shortcomings, so that it could move on to the next stage: factory trials.

Meanwhile the aircraft was given the military designation Yak-15 and factory designation Yak-15RD-10 or Yak-RD in short.

According to the plan, factory trials commenced on 9 April 1946 and were going to last until July. Yak-15RD-10 flew for the first time on 24 April - notably the competing I-300 (Product F) from Mikoyan and Gurevich OKB-155 also flew for the first time the same day. (Allegedly a coin was tossed to decide which prototype would fly first; the MiG team won.) During the next flight on 23 May, with a take-off weight of 2,570 kg (5,775 lb), speeds of 770 km/h (481 mph) near the ground and 805 km/h (503 mph) at an altitude of 5,000 m (16,250 ft) were achieved. With fuel tankage of 472 kg (1,061lb) the range was 575 km (359 miles), and time to of climb to an altitude of 5,000 m was 4.1 minutes. It was found that when the aircraft exceeded 6,000 m (19,500 ft) the engine was unable to increase revs. Final state trials were completed on 22 July 1946. Soon Yakovlev was able to file a report to his superiors that his mission had been accomplished.

All performance and flight characteristics were analysed and officially approved. The positive trial results prompted the OKB to display the aircraft in public. Having obtained the necessary permission the prototype was demonstrated on 18 August during the Soviet Aviation Day at Tushino. The competing design from Mikoyan – I-300 (MiG-9) - was also displayed during this air show.

This was an obvious success by these two OKBs, with the other teams left far behind. The day after the Tushino display both chief designers were invited

Side views of Yak-15RD10. 1/72 scale.

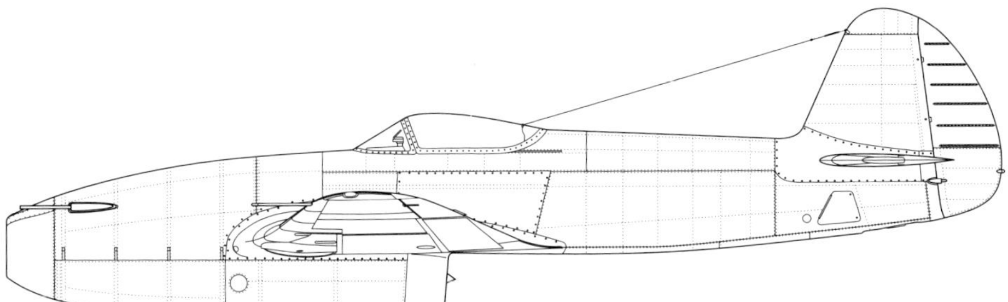

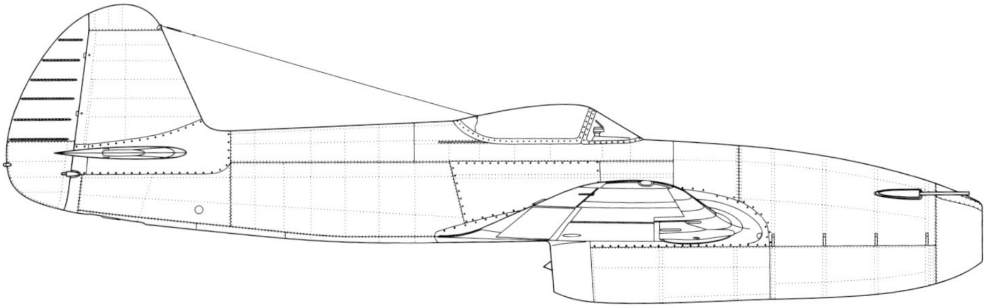

to the Kremlin, where they heard much praise during a meeting with Joseph Stalin, but they were also told that 15 of each design should be demonstrated during the flypast on 7 November, the October Revolution Day. To allow this task to be accomplished, on 12 September 1946 the Ministry of Aircraft Production issued an order to manufacture small batches of these aircraft, and to earmark two of each for trials at the *GK NII VVS Zhukovski*. This special "flypast" batch of Yak-15s would be completed without armament, with limited radio equipment and with an enlarged fuel tank (in place of armament). Clearly, this was due to the insufficient range. Production of the batch was allocated to No. 31 Plant at Tbilisi, which at the time still assembled Yak-3 fighters. The task was rather complicated for the plant. It had to build from scratch, in a short time and without proper experience, tooling or equipment, a batch of fifteen jet-powered fighters. Despite the problems the crew of the plant did their job, working in three shifts. Their job was facilitated by the fact that complete Yak-3 fuselages were available on the production lines, and these were used for production of new Yak-15RD-10 aircraft. The first batch of RD-10 engines to be fitted in the 15 Yak-15s suffered from extensive problems, as some components were still of German origin. For this reason the Ufa plant suggested a low operational life of 25 hours. To prepare the aircraft for the display they were ferried to the assembly airfield. On 7 November all plans were ruined by the weather, and the parade was held without any flypast. The entire effort was for nothing and no propaganda effect was gained.

Meanwhile two aircraft of the first batch (factory nos. 310002 and 310015, coded "32" and "35", respectively) were handed over for trials at Zhukovski, and state trials were completed there during December 1946 – April 1947. The aircraft were armed with just a single NS-23 cannon and had a take-off

Side views of Yak-15, c/n 31015. 1/72 scale.

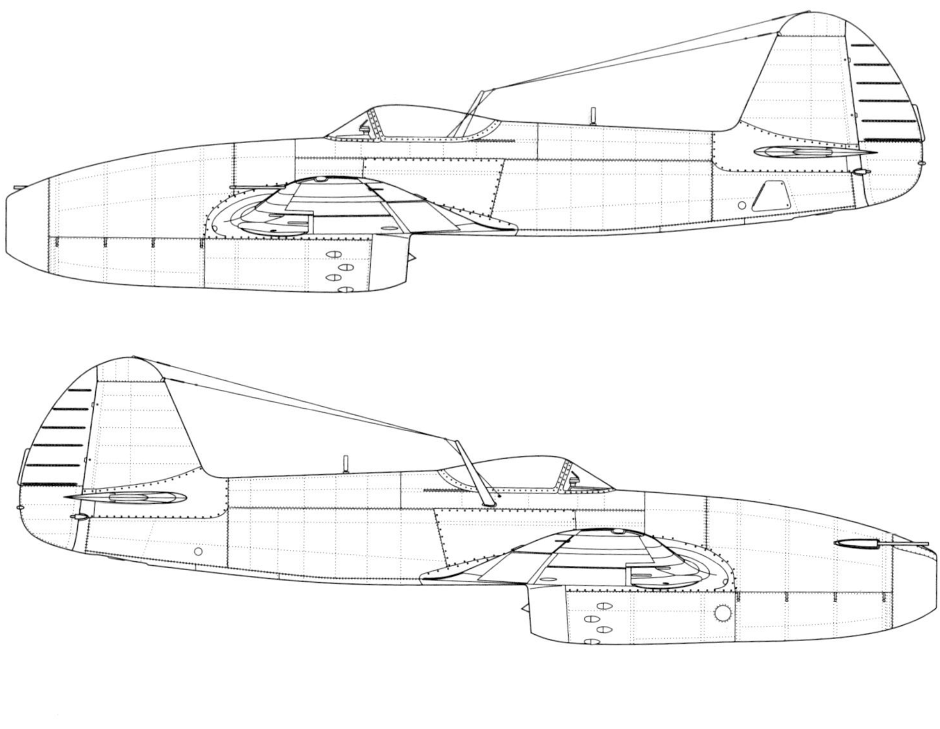

weight of 2,742 kg (6,162 lb). Two test pilots, Grigoryi A. Sedov and Afanasyi G. Proshakov, were named to carry out flying trials. Test results proved slightly lower than expected, and the aircraft received a number of limitations, including the maximum speed of 700 km/h (438 mph) at an altitude of up to 3,150 m (10,238 ft) and of Ma 0.68 at higher altitude, thus reducing its specifications. The limitations resulted from the poor power plant which caused a lot of trouble in operation.

Yak-15 c/n 31015, yellow 35, the 15th production aircraft during State acceptance trials.
Stratus coll.

Side views of Yak-15, c/n 31002. 1/72 scale.

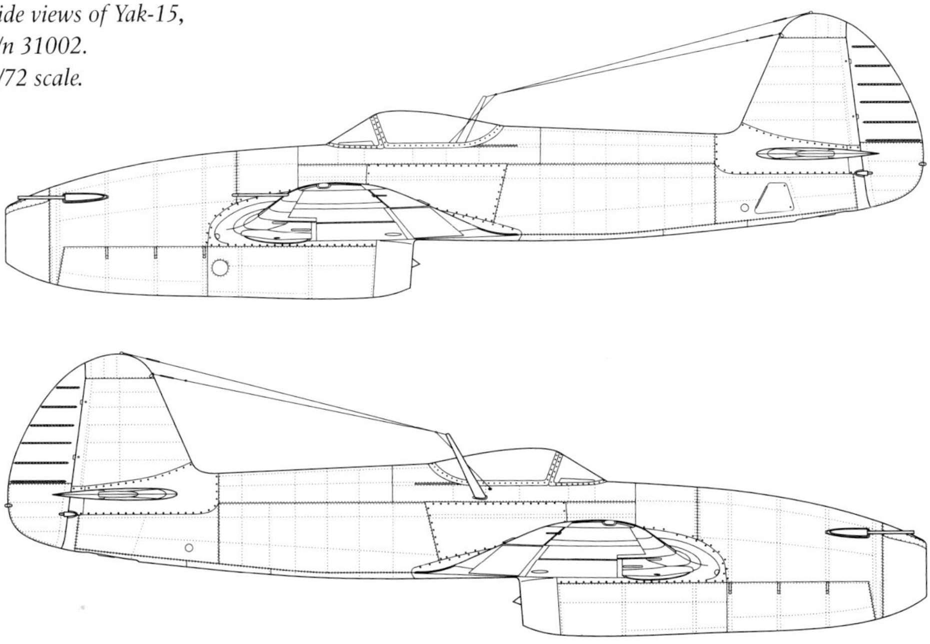

Military trials also demonstrated a number of faults and shortcomings, needing to be cured. A lot of criticism was made of the undercarriage layout, as during taxiing exhaust gases engulfed the tail wheel, and the latter was often damaged by objects torn up from the airfield surface. Visibility from the cockpit during taxiing was also limited. Example no. 310009 was used for piloting trials. The test pilot, Petr M. Stefanovskiy, considered the aircraft easy to fly, the best of its generation of Soviet jet fighters. It could be flown by pilots of medium skills, a significant convenience in the training process.

In this situation a decision had to be taken as to the further fate of the design. On 16 December 1946 the Council of Ministers issued directive no. 2698-1114 which ordered the manager of No. 31 Plant to manufacture another batch, of 50 examples, including 25 examples of the single seat fighter version and 25 training two-seaters. Production plans called for 5 Yak-15s to be built in January, 15 in February, 20 in March, and 10 in April. Production of the batch was also linked with a propaganda effort planned for 1 May 1947.

As ordered by the Ministry of Aircraft Production all aircraft of the second batch (25 single seaters and 25 2-seaters) were built according to the pattern of the first batch: without armament (NS-23K cannon, ammunition) or full radio-electronic equipment (RPKO-10M radio-direction finder). The plant completed the order by the end of April.

On 1 May 1947 the weather was good and 50 each Yak-15s and MiG-9s took off for the flypast. This time the display was impressive, with 100 jet-powered aircraft flying over Red Square. The aircraft were grouped in a column of 5-aircraft vics. A total of 20 vics flew past at short intervals. The effect was excellent, the rulers were satisfied, and propaganda was triumphant. Few people realised that the short range of the aircraft forced some limitations upon the organisers of the flypast. The aircraft had to take off in groups of five at short intervals, as assembling them after individual take-offs would have taken some

Initial production Yak-15. ***Yakovlev OKB photo via author.***

Initial production Yak-15 at static display.

time. The airfield most conveniently located in terms of distance to be flown was selected for the formation.

In May 1947 the GK NII VVS issued a document that declared the aircraft as fulfilling the air force specifications for a jet trainer. At the same time it was ordered to develop a two-seater combat trainer with dual controls and tricycle undercarriage.

Rapid progress in jet-powered aircraft design in Western Europe and the USA continued to force the Soviet authorities to develop new, higher and more demanding specifications. On 11 March 1947 design offices received a government instruction to build fighter aircraft with new power plants of higher thrust, able to achieve a maximum speed of 1,000 km/h (620 mph). Following the May display the air force prepared for the annual Aviation Day, culminating in the central ceremonies at Moscow-Tushino aerodrome. Principal items during the air show would include Col. I. Polunin's aerobatic display in a Yak-15 and a display of three Yak-15s led by Col. Nikolay I. Khramov. The section would perform a number of vertical and horizontal aerobatic figures in vic formation.

Yak-15s were allocated to fighter aviation regiments where they were used as transition aircraft. Although they were recommended as a training type, during two years (1946-1947) No. 31 Plant at Tbilisi built 280 examples that were delivered to combat fighter aviation regiments based in the USSR, Poland, Germany, Hungary, Romania and Manchuria (9th Air Army). Aircraft factory numbers consisted of six digits, such as 310009, where the first two digits "31" denoted the factory, the next two "00" the batch number and the last two "09" the individual aircraft number. This was one of two jet fighter aircraft types, the other being MiG-9, introduced into the Soviet Air Force. The MiG-9 had slightly better performance and was built in larger numbers (400), allocated to regiments of 1st Air Army in Belarus, 7th Air Army in Azerbaijan, 14th Air

Army in Ukraine, 15th Air Army in the Kaliningrad area and 16th Air Army in East Germany.

PARAMETER	Yak-15	MiG-9	ADVANTAGE
Engine (thrust)(kg)	1 x RD-10 / 900	2 x RD-20 / 2 x 800	MiG-9
Empty weight (kg)	1,852	3,420	Yak-15
Take-off weight (kg)	2,742	5,054	Yak-15
Speed at sea level (kph)	700	864	MiG-9
Speed at 5,000 m (kph)	786	910	MiG-9
Rate of climb m/s	20		
Service ceiling (m)	13,350	13,500	MiG-9
Range (km)	510	800	MiG-9
Length/Span (m)	8.7 / 9.20	9.83 / 10.0	
Wing area (m^2)	14,85	19,2	
Take-off run (m)	600	950	Yak-15
Landing run (m)	530	1,060	Yak-15
Armament	1 x 23 mm cannon	1 x 37 mm cannon 2 x 23 mm cannon	MiG-9
Number of rounds	60	40 2 x 80	MiG-9

Yak-15 in flight. ***Yakovlev OKB photo via author.***

A well known photo of a production Yak-15in flight.

Yakovlev OKB photo via author.

Of the 11 representative characteristics seven were in favour of the MiG-9. Additionally, the layout of the MiG-9 was much better than that of the Yak-15. The tricycle undercarriage gave it improved forward visibility from the cockpit. Without a doubt the armament of the MiG-9 was significantly better, consisting of three cannon, including a 37 mm one. The Yak-15 had lower weight, it was easier and simpler to fly. Pilots were able to quickly convert onto these machines, this being a significantly favourable feature. Yak-15s were displayed at flypasts and air shows many times. In 1948 a display team of five machines was formed under Gen. Evgenyi Y. Savitsky. The other pilots were Petr S. Sereda, N. I. Khramov, Vasily V. Yefremov and P. Solovyev. Another well-known team of five Yak-15s during that period was led by Hero of the Soviet Union Col. Pavel F. Chupikov.

During operations in combat units the aircraft displayed some shortcomings, most often failures of the main undercarriage shock absorbers and of the tail wheel shock absorber. The short endurance was a major disadvantage. During training the aircraft could not depart from the airfield over longer distances (circuits, cruises). When fully fuelled (590 kg [1,326 lb] of kerosene) the aircraft could fly just over 500 km (312 miles). With provision for navigation reserve and strong winds, for example, the aircraft was only able to operate within a hundred kilometres of its base, with an endurance not exceeding 25-30 minutes. Attempts were made to cure the problem in various ways, including in-flight refuelling.

In 1948 Vladimir S. Vakhmistrov was ordered to develop an air-to-air refuelling system. Following the pattern of Flight Refuelling Limited in Britain, he used the flexible hose method. The tanker and refuelling aircraft deployed hoses and manoeuvred to join them. Fuel was then transferred by gravity. The system had many faults and was not used in practice. The idea was significantly improved and brought to the flying trial stage by two pilots, Igor I. Shelest and V. Vasyatin. A Tu-2 bomber was used as tanker aircraft during the trials, with Yak-15 "37" as receiving aircraft. Trials continued successfully for some years, eventually leading to a practical system.

Yak-15 'Yellow 47" was used for trials with in-flight refuelling system.
Yakovlev OKB photo via author.

Below: Yak-15 "20"
1/72 scale.

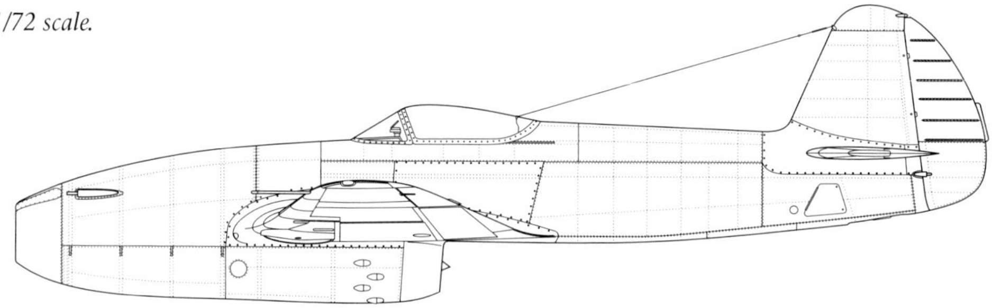

Below: Yak-15 "32"
1/72 scale.

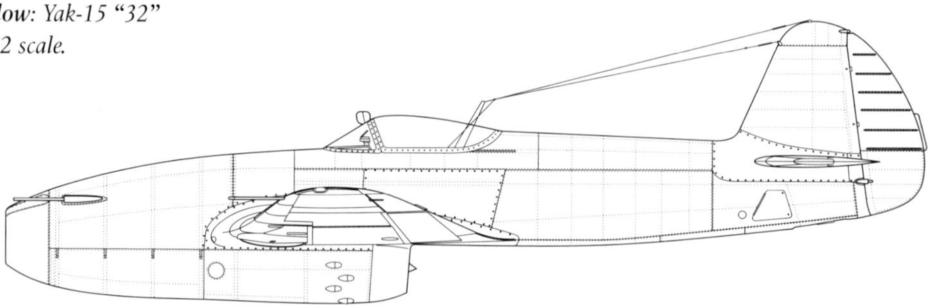

Below: Early production Yak-15.
1/72 scale.

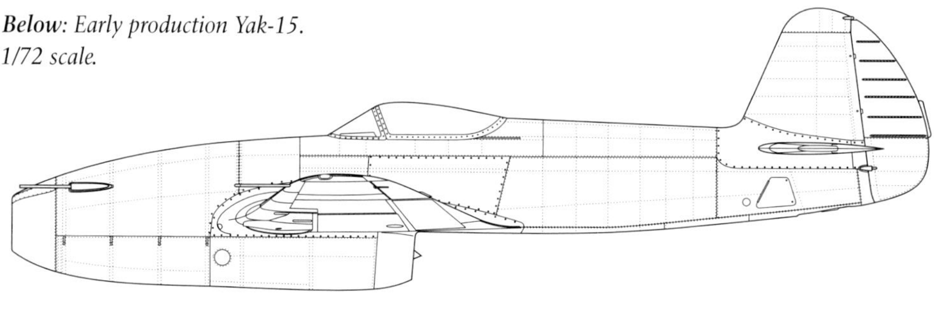

Below: *Yak-15 "37" without armament.*
1/72 scale.

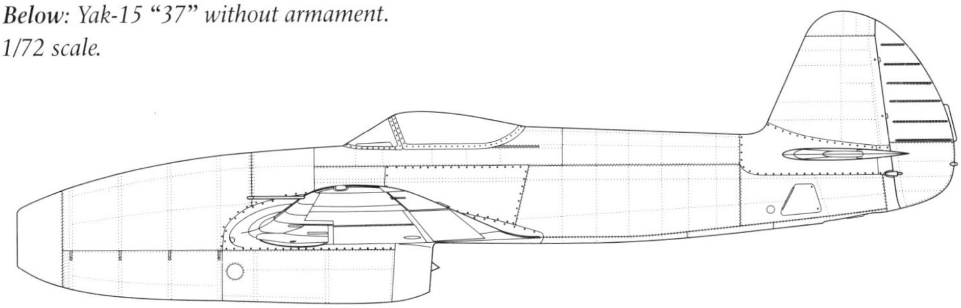

Below: *Yak-15 "37" with armament installed.*
1/72 scale.

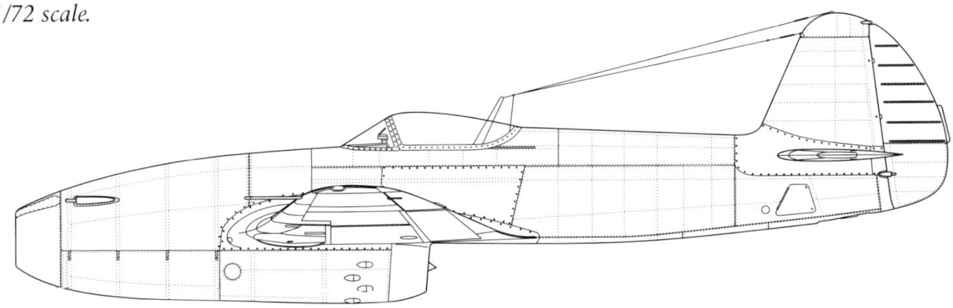

Production Yak-15 during take-off.
Yakovlev OKB photo via author.

Yak-15.
1/72 scale.

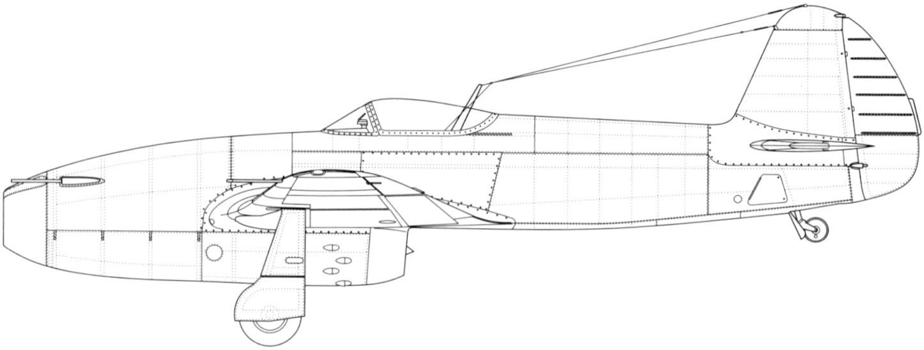

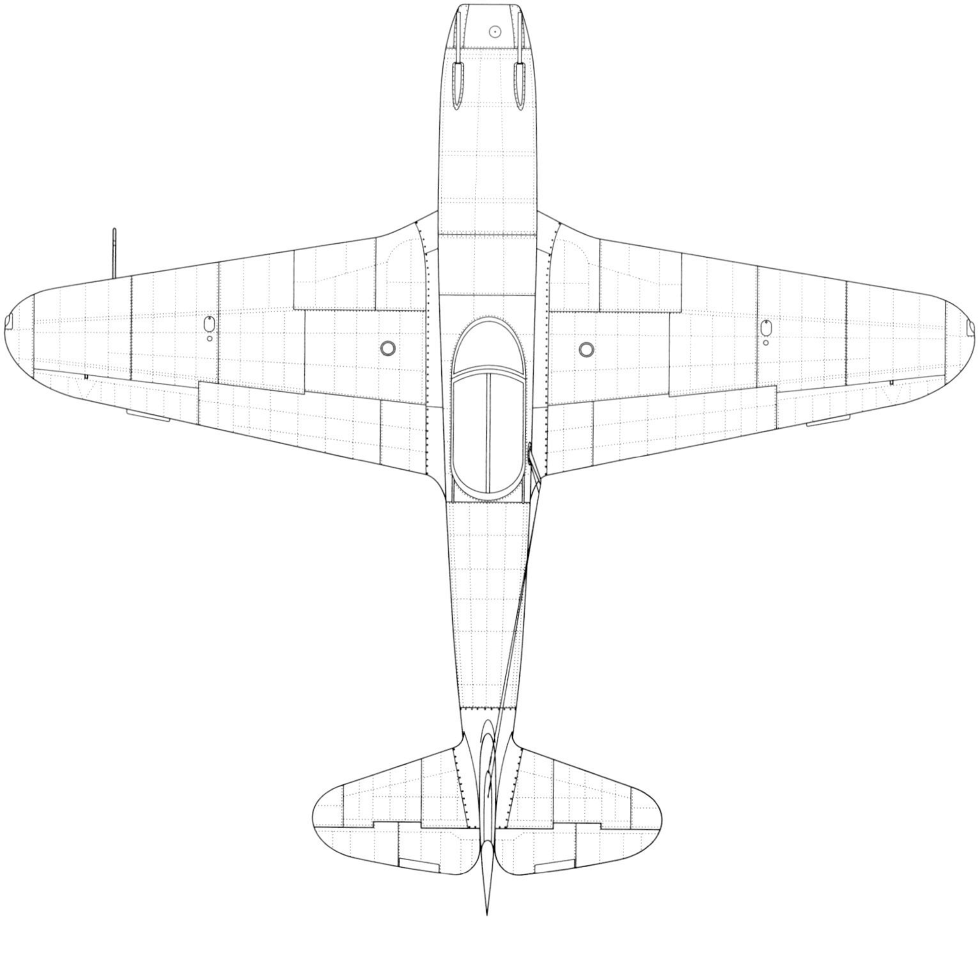

Yak-15.
1/72 scale.

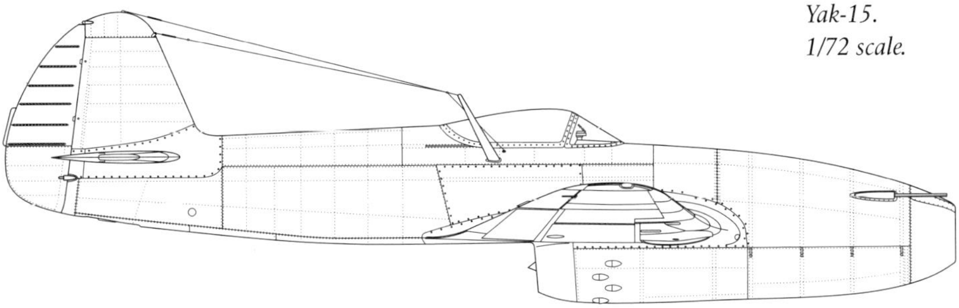

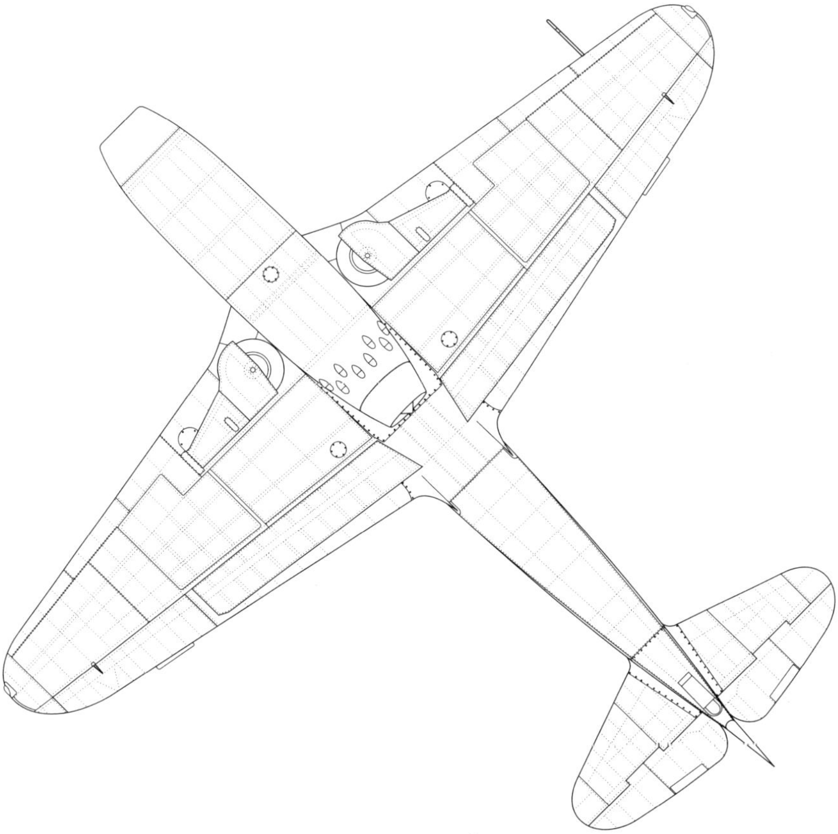

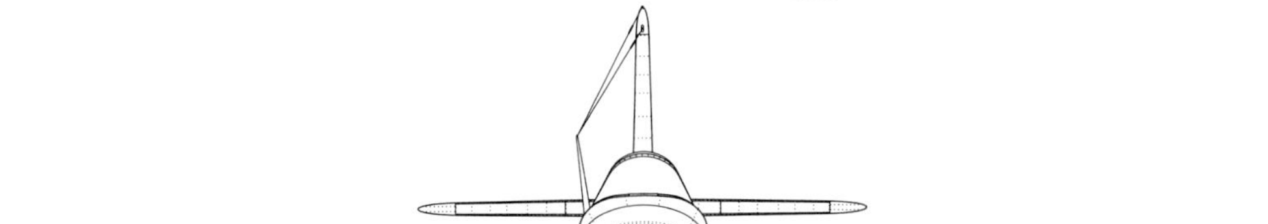

Birth of the Yak-17

During two years the Soviet Air Force took delivery of some 680 jet fighters, that equipped seven air divisions located in strategic areas of the USSR and its allies. The importance and urgency of work on new jet-powered aircraft designs is best shown by the fact that three days after the first flight of the Yak-15, the USSR Council of Ministers issued directive no. 952-397 which gave the OKB-115 headed by A. S. Yakovlev the task of developing a new fighter with improved characteristics and performance. The aircraft was going to be powered by the RD-10 engine and to have a significantly improved aerodynamic layout. It was supposed to reach a speed of 850 km/h (527 mph) at an altitude of 5,000 m (16,250 ft) and a range of 700 km (438 miles). The project retained the general layout of the Yak-15, but all inadequate or limiting systems were changed. The wing was redesigned, with laminar-flow airfoils developed by TsAGI. The undercarriage was changed to a tricycle layout and the main wheels were enlarged. An ejection seat was introduced, and an armour glass plate was fitted in the windscreen. The new aircraft was going to be ready in the autumn of 1946.

Slightly earlier, in June, the Council of Ministers issued another directive, no. 1498-665, which led to order no. 525 of the Ministry of Aircraft Production dated 6 August, that called for development of a two-seat trainer for conversion training of air crews onto jet-powered aircraft. The aircraft was to be fitted with the Jumo 004 engine. The first prototype was to be ready by 1 November, to be sent for trials on that date. The design bureau named the project the Yak Jumo 004 Vyvoznoy (Conversion Trainer; also known as the Yak-15V), and its military designation was Yak-21.

Meanwhile, work on the new fighter continued at the planned pace. The first example was completed on 3 September 1946, 6 months after the first flight of the Yak-15. The aircraft received the military designation Yak-17. On 26 September OKB-115 test pilot Georgiy S. Klimushkin performed several taxi runs, and on 4 October he took off for the first time. Subsequent flights revealed a number of shortcomings and design faults. During one of the flights failure of the new undercarriage led to damage to the rear fuselage. Following this accident the aircraft was sent back to the manufacturers for repair and modification. Work continued until 15 March 1947.

Since No. 31 Plant's production capability was being fully used, construction of the prototype Yak-21T was entrusted to No. 464 Plant at Dolgoprudniy. Construction of the Yak-21T prototype (factory no. 01464, code "22") was

Yak-17-RD10 development version.

J. Gordon coll. via author.

Yak-21 prototype, "Yellow 101".

Yakovlev OKB photo via author.

completed in April, and in May the first flight took place (test pilot G. S. Klimushkin) followed by factory trials under leading engineer Nikolay I. Leonov. It was designated by the air force as the Yak-17UTI (Uchebno-Trenirovochniy Istrebityel - Primary/Advanced Training Fighter). Between 31 May and 6 August state trials were completed successfully at Zhukovski, and despite some shortcomings the aircraft was recommended on 17 September 1947 by the GK NII VVS at Zhukovski for series production at Tbilisi. The trials proved that the Yak-21T/UTI Yak-17 had slightly poorer performance than required (maximum speed and rate of climb). Some other faults were found, including the fact that the instructor was not able to start the engine, or retract and lower the undercarriage and flaps. Despite these shortcomings the pilots who ran the state trials declared that the aircraft could fly at high angles of attack, was easy to recover from a spin, and could easily endure 8G forces in aerobatics. It was suggested that production examples be fitted with a single NS-23 cannon with S-13 camera gun. Series production of the Yak-17UTI was located at No. 31 Plant which started assembling them in the autumn of 1947.

Test pilots stressed that the introduction of the new undercarriage layout in the Yak-17UTI significantly improved handling of the aircraft during taxiing, take-off and landing. This improvement and many others were used in the Yak-17 combat version. The nose wheel was retracted into an additional fairing situated in the lower forward fuselage. Main wheels were repositioned to the rear, so the wing had to be redesigned. The changes resulted in increased fuel

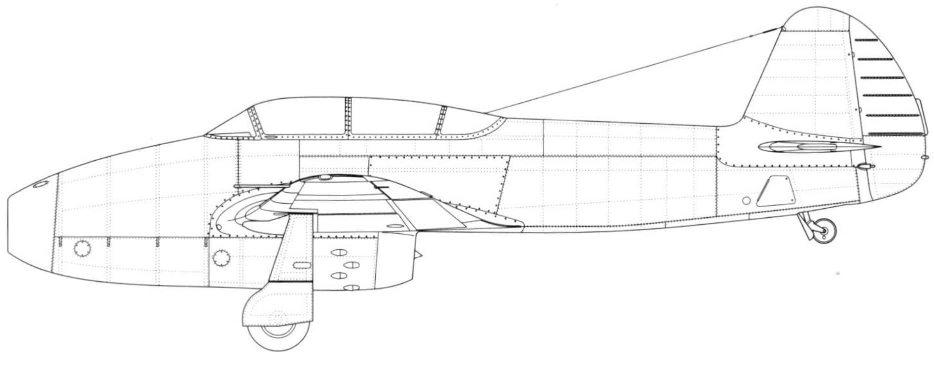

Yak-21. 1/72 scale.

Yak-15U-RD10 prototype during state acceptance trials.
Yakovlev OKB photo via author.

capacity from 420 to 460 litres. To increase range, attachments for two wing tip 200 litre tanks were also fitted. The PBP-1A optical gun sight was replaced with the ASP-1 gyroscopic gun sight, and the PAU-22 camera gun was fitted in the starboard wing.

The prototype of the Yak-17U single-seat fighter was built by No. 464 Plant, with the Yak-15 factory no. 310002. This example was given the military designation Yak-17. Following ground trials, the aircraft was prepared for the first flight. The Yak-17 first flew on 6 June 1947 with the test pilot G. S. Klimushkin at the controls. The first flight was successful and factory trials were conducted under V. I. Yemelyanov, completed on 10 August. The aircraft was displayed during the Aviation Day on 3 August at Tushino. The next stage, the state trials, were completed between 27 August and 15 November. G. A. Sedov was in charge of the trials that were conducted by pilots P. M. Stefanovskiy, Andrey G. Kochetkov, A. G. Proshakov, Alexandr P. Suprun and others. Trial results proved that the aircraft had better performance than its predecessor, the Yak-15. Its take-off weight was 2,890 kg (6,495 lb), and with external tanks 3,240 kg (7,280 lb). In this form its range was 717 km (448 miles). The tricycle undercarriage largely

Yak-15U-RD10 prototype with drop tanks attached, during state acceptance trials.
Yakovlev OKB photo via author.

Side view of Yak-15U-RD10 prototype during state acceptance trials.
Yakovlev OKB photo via author.

simplified take-off and landing. When flying with external tanks the aircraft showed a tendency to vibrate around its vertical axis. Despite these comments the committee approved the aircraft's performance. Based on the directive of the Council of Ministers no. 878-280 of 20 March 1948, the aircraft was placed in series production at No. 31 Plant as the Yak-17. Even during its trials preparations for series production were undertaken at the plant. More alterations were introduced in the Yak-17 design during series production. To improve the aerodynamic layout and controllability, the tailplane span was increased from 3.25 m (10.56 ft) to 3.60 m (11.7 ft), the modified NS-23K cannon was fitted, and an S-13 camera gun was fitted on the windscreen.

In early March 1948, with series production underway, a selected example was sent for checking trials that were completed successfully. Based on the report of the trials, the USSR Council of Ministers issued directive no. 878-280 confirming series production.

During production 15 examples of the 11th and 12th production batches were fitted with the more modern ASP-3M gun sight, and from no. 311222 on USB-1 ammunition counters were introduced. The last batches were also fitted

Front view of Yak-15U-RD10 prototype.
Yakovlev OKB photo via author.

with the modified RD-10A engines with the operational life between inspections increased from 25 to 50 hours.

By the end of 1948 the factory had assembled 279 examples. During the next and final year of production 151 Yak-17s and Yak-17UTI s were built. Altogether the Tbilisi factory turned out 430 Yak-17s of both versions. It is probable that 280 were built as fighters and 150 as trainers. Yak-17s were given six-digit factory numbers from 310001, while Yak-17UTIs had longer, seven-digit numbers, with an extra digit introduced to designate the type of aircraft. For example, no. 3120177 meant: 31 – the factory, 2 – aircraft type, 0177 – individual aircraft number.

A production Yak-17 was a single-seat, all-metal low wing monoplane with tricycle retractable undercarriage. The fuselage was built around a trusswork welded from steel tubes, with streamlined fairings on top and bottom. Fuselage skin of duralumin was fitted with numerous inspection openings and removable panels. Bottom fuselage aft of the engine exhaust nozzle was covered with heat resistant steel. The all-metal two-spar wing with tapered planform and rounded tips had stressed skin. Modified Clark-YH airfoil was used. An FS-155 landing light was fitted in the port wing. The wings were fitted with Frise-type ailerons and split landing flaps. Cantilever tail surfaces were unswept, with the rudder covered with reinforced fabric. Elevators were fitted with trim tabs. The nose wheel leg was retracted backwards into a streamlined fairing. 600x180 mm main wheels were fitted with pneumatic brakes and oleo shock-absorbers, and retracted into wells in the wings.

The aircraft was powered by the RD-10A engine that weighed 780 kg (1,753 lb), with eight-stage axial compressor, single-stage turbine and six combustion chambers. It had a static thrust of 9.8 kN (1,000 kG) at 8,760 rpm. The engine was fitted at 4° to the aircraft centreline. The power plant was started by a two-stroke piston engine. The fuel system consisted of seven tanks, six in the wings and one 240 l tank in the fuselage. Additionally it was possible to attach two 200 l tanks under the wing tips. The cockpit canopy was two-piece, with a fixed windscreen with armoured panel of bonded Plexiglas, and a sliding hood covered with 6 mm Plexiglas. The sliding hood was fitted with an emergency jettison system. The lightweight ejection seat could be vertically adjusted by 65 mm, depending on the height of the pilot.

The pneumatic system was used for lowering and retracting the undercarriage and flaps, operating the main wheel brakes and reloading the cannon. The main system was duplicated with an emergency one. Air at 50 atm was held in two 12 litre spherical containers. The oxygen system consisted of a KP-14 oxygen

Production Yak-17s waiting for take-off for the parade in Moscow.
Author coll.

Preserved Yak-17 at Monino Museum.
M. Czaplicki.

regulator and a 4 litre cylinder. Electric power was supplied from a GSK-1500 alternator (alternating current) and from a 12A-10 battery (direct current). Radio equipment included the RSI-6K transmitter-receiver, RPKO-10M radio-direction finder and SCh-3M IFF. A complete set of flying and navigation instruments was located on the panel in the cockpit (KUS-46 airspeed indicator, VR-30 rate of climb indicator, "Horn" artificial horizon, KI-11 compass, M-46 Mach indicator and AVR-M time clock). Late batch examples were additionally fitted with a PDK-46 remote compass. Armament consisted of two fast-firing NS-23 cannon with 60 rounds per gun.

The Yak-17UTI training version had a similar layout to that of the combat version, the principal difference lying in the two-seat cockpit with the instructor's seat at the rear and the pupil's at the front. One of the fuselage-mounted fuel tank had to be discarded to make room for the enlarged cockpit. This reduced the aircraft's maximum range from 717 km (448 miles) to 370 km (231 miles). To allow in-flight communication between crew members, the SPU-2M intercom was fitted. The canopy was made of five pieces, including three fixed and two rearward sliding hoods. The aircraft had no armament, just the camera gun for gunnery training. The principal shortcoming, that significantly reduced its training capability, was its short endurance of 30 minutes.

Naturally, the Soviet Air Force was the main recipient of production Yak-17s. The first examples went to units in the second half of 1948. The conversion process of individual air regiments continued until the end of 1949, when the latest MiG-15 swept-wing fighter entered service. Some Yak-17UTIs went to regiments equipped with Yak-15s to accelerate conversion and training of pilots. Combat versions were introduced in subsequent regiments, significantly increasing the potential of the air force. The ease of flying the aircraft is best

Production Yak-17UTI.
Yakovlev OKB photo via author.

shown by the fact that during the Aviation Day air show in 1949 an aerobatic team of nine Yak-17s performed, led by Col. Nikolay N. Shulzhenko.

The same feature meant that conversion training on the type was rapid and safe. As an example, the newly formed 472nd Fighter Aviation Regiment at Kursk completed conversion training of the entire personnel onto Yak-17s between March and July 1950. The aircraft, as one of the first jet designs in the USSR, suffered from a number of shortcomings and faults that affected its combat capabilities. It did not offer significant combat potential, even more so as its operational reliability was poor, especially in the case of the power plant, which suffered from surging, fires and uncontrolled start-up. Starting it up in the air was complicated and consisted of several stages: 1 - to start the start-up petrol engine, 2 – to start the compressor, 3 – to start the main engine on light fuel, 4 – to switch to kerosene. This, and other characteristics, meant the aircraft

Yak-17UTI (Yak-17 W) of Eskadra Pilotażu Myśliwskiego *(Fighter Pilots Squadron), Polish Air Force at Tomaszów Mazowiecki airfield. Aircraft's technician ppor. Franciszek Dziekan to the right.*
F. Dziekan

Yak-17 at Soviet airfield, late '40s.

Authors coll.

was regarded as a transition fighter, for conversion training of pilots onto the new class of machines, hence its limited series production.

During that period (1947-1949) the Soviet Air Force used some 1,100 jet-powered fighter aircraft of various types. The design rivalry was won by Alexandr Yakovlev, whose designs constituted a majority (64%) of this number. In the late 1940s and early 1950s the situation started to change rapidly, with mass production of the revolutionary MiG-15. This aircraft quickly phased out the older and inferior Yak-15s and Yak-17s. The latter were offered to selected Soviet satellite countries, where they entered service in small batches.

Side views of Yak-17UTI. 1/72 scale.

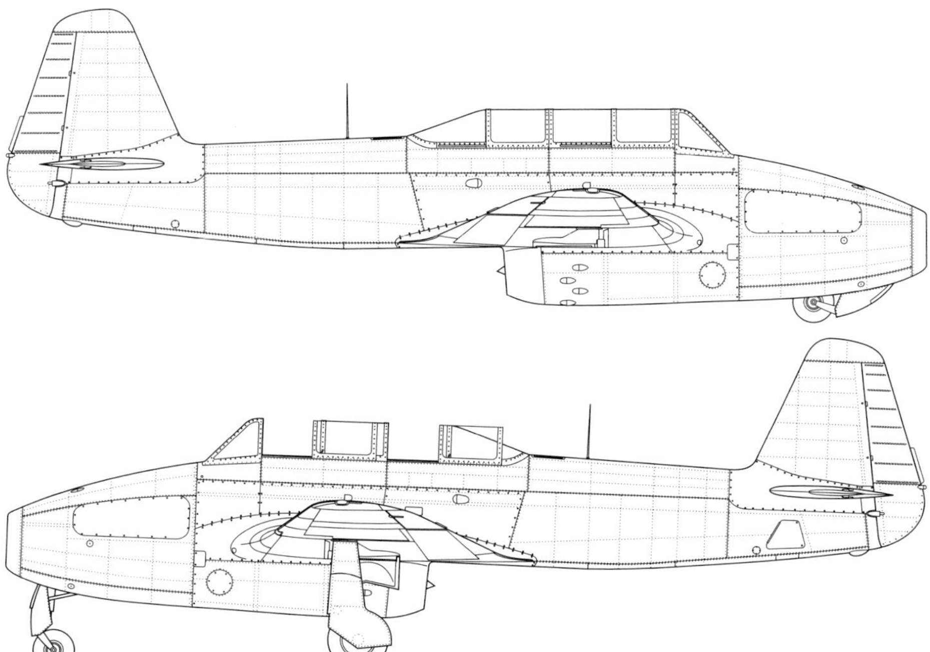

Yak-17.
1/72 scale.

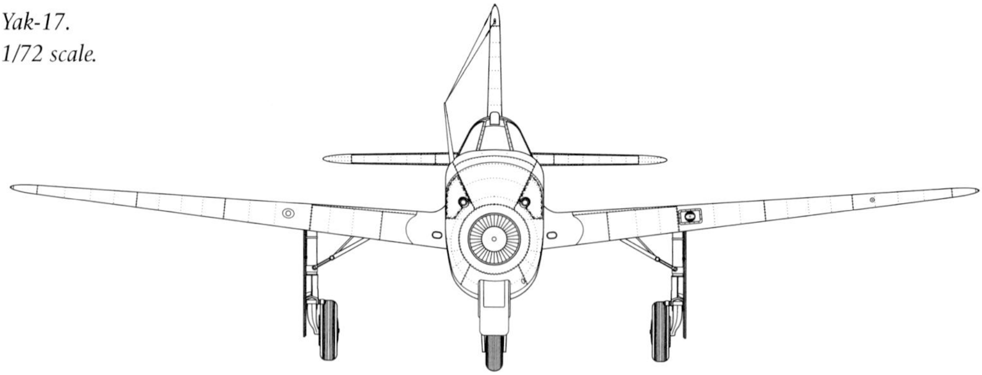

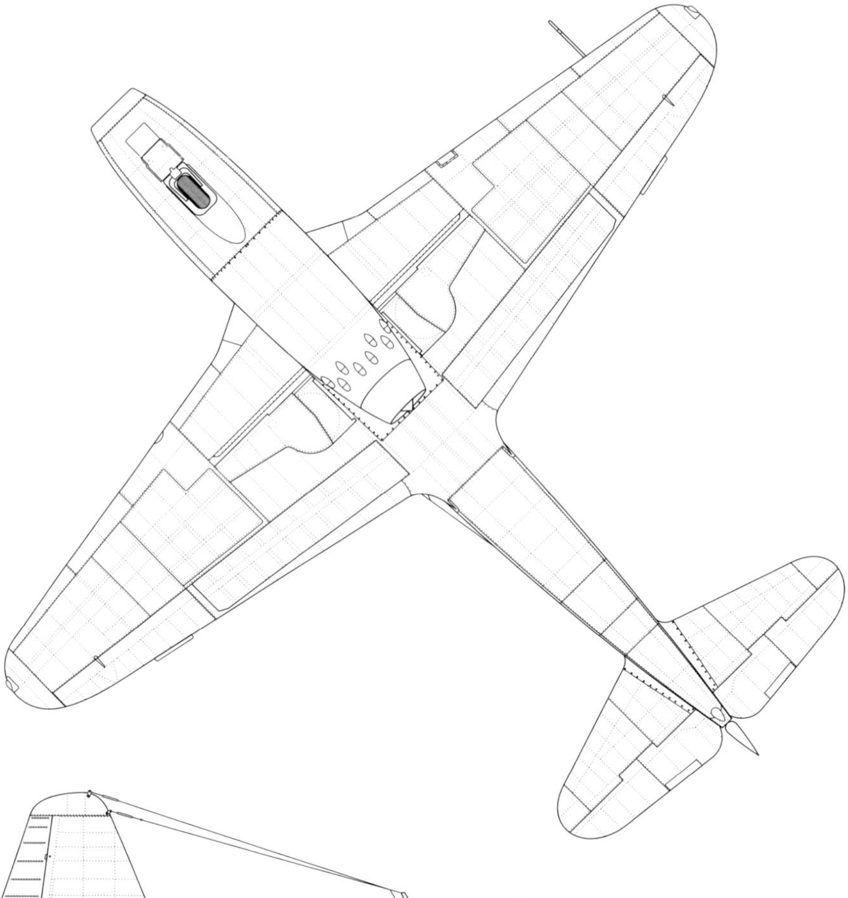

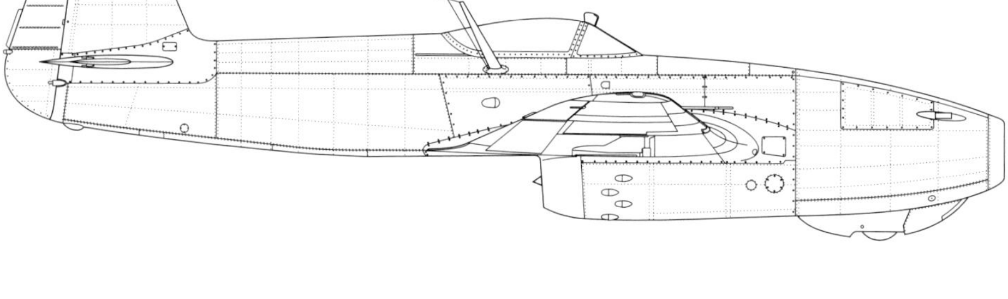

Yak-17.
1/72 scale.

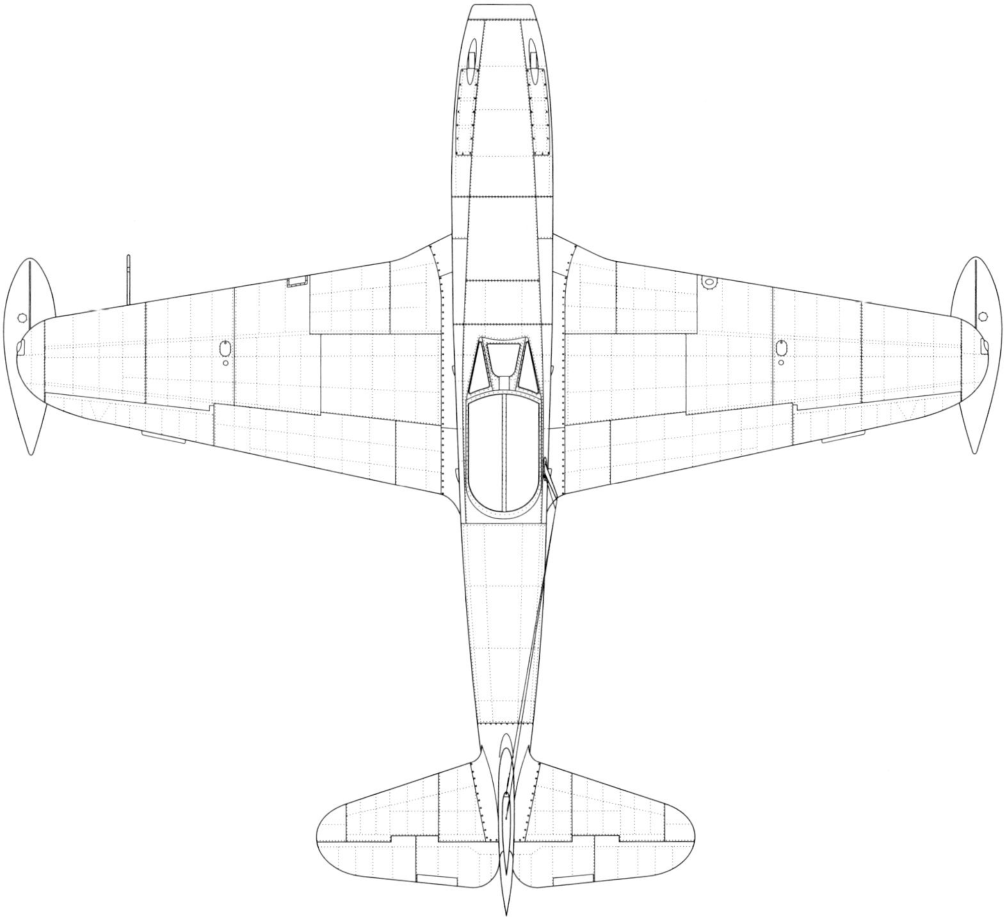

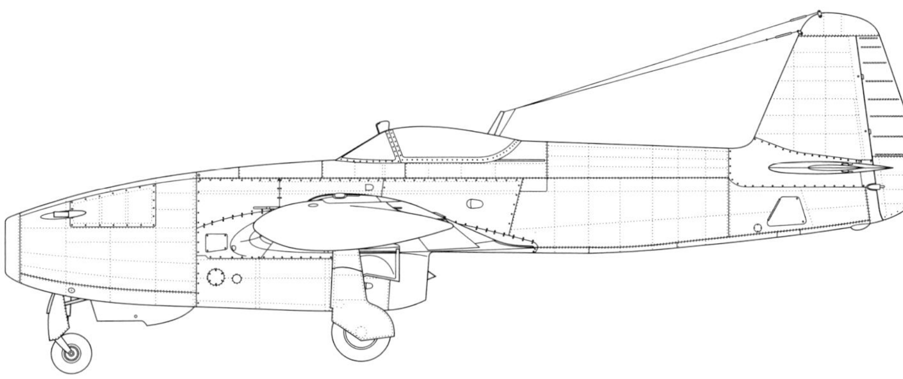

Yak-17 successors

The relatively low number of Yak-17 and Yak-17UTI aircraft built was due to rapid progress in developing a new design. At the time the western powers continued to be at the forefront of technological progress, introducing more and more modern solutions, and their aircraft showed better and better characteristics. Work in the USSR continued extensively, but lack of suitable power plants prevented rapid progress. Captured Jumo engines did not have enough thrust to meet the requirements of the authorities. To find a solution the central authorities organised an industrial-military meeting at the Kremlin as early as April 1945, focused on development of prospective jet-powered combat aircraft. It was decided that the engines provided the starting point and that work on these in the USSR advanced too slowly. The only rapid solution was to purchase such engines in the West. The result was rather unexpected: in September 1946 Britain sold a batch of Nene and Derwent 5 engines to Soviet representatives.

Selected design offices started working on promising jet-powered aircraft using these new engines. OKB-155 was already working on the successor to the MiG-9. A new fuselage had to be designed, to house one of the newly acquired engines that featured a centrifugal compressor and therefore was of much larger diameter. The new concept was a straight fuselage of circular cross-section with the engine at the rear, the air intake at the front with the duct split around the cockpit. The swept wing and tail surfaces were an innovation as well. The design received the designation I-310 (*Istrebityel* - Fighter). Meanwhile No. 45 Plant developed a copy of the Nene under the designation RD-45. In 1947 three pro-

The first prototype of the Yak-19. "Yellow 24", c/n 01115001, in flight. ***Stratus coll.***

totypes of the new fighter were built, one of them first flown on 30 December. Already during initial flights the design proved a revelation. In the spring of 1948 a decision was taken at central level to start mass production.

At the same time the OKB-115 designers under A. Yakovlev worked on a new fighter in parallel in three teams.

The first concept was a lightweight fighter aircraft powered by the 12.35 kN TR-1 engine locally developed by OKB-165. It was much larger than the RD-10, so the fuselage was suitably redesigned. It was oval in cross-section, and the power plant was located in the rear fuselage. Straight wings were adopted from the Yak-17. The cockpit was located in the forward fuselage in the middle of the air intake. This was the first Soviet jet fighter with such a layout. Since work on the TR-1 engine was delayed, it was decided to use the RD-10F with afterburner. The project was designated the Yak-19. Construction of the prototype was completed on 29 November 1946. Factory trials continued from 10 December 1946 to 27 September 1947, and the first flight was made on 8 January 1947 (test pilot M. I. Ivanov). Half a year later the second prototype Yak-19 was first flown, this being fitted with two additional tanks at the wing tips. The second prototype underwent state trials between 17 October 1947 and 30 January 1948. However, constant problems with the afterburner resulted in a negative opinion of the aircraft.

Another team of designers from Yakovlev OKB finished work on another fighter on 1 February 1947. Leonid L. Selyakov was appointed the leading designer. The general design layout followed that of the Yak-19: fuselage with the central air intake at the front and engine at the rear. Straight wings were slightly larger than in the Yak-19, with a thinner laminar flow airfoil. Both tail surfaces were swept. The pressurised cockpit at the front was fitted with an ejection seat. Construction of the first prototype, designated the Yak-25-1, was completed on 29 October 1947 and the following day it commenced factory trials that continued until 3 July 1948. It was first flown on 2 November 1947 (test pilot Sergey N. Anokhin). The second prototype, Yak-25-2 built in parallel, was earmarked for state trials, conducted between 23 July and 9 September 1948.

The second prototype Yak-25, c/n 115002.
Stratus coll.

Yak-23 development

Another team of designers worked on the personal concept of the chief designer who insisted on continuing the line of the Yak-15 and Yak-17. The project was a lightweight highly manoeuvrable battlefield fighter. The aircraft was going to be fitted with a power plant based on the Derwent 5 (RD-500). The aerodynamic layout was inherited from its predecessors, but the structure was designed anew as an all-metal semi-monococque. The aircraft was a lightweight mid-wing monoplane with unswept wing and tail. The fuselage, with a stepped layout, had the engine in the forward part. The cockpit, located in the middle of the fuselage, had a sliding hood and was fitted with a lightweight ejection seat. To reduce the empty weight no cockpit pressurisation or air brakes were fitted, while armour and fuel tank self-sealing was reduced. Wing skin thickness was reduced as far as possible. As a result a record low weight was achieved: 1,902 kg (4,725 lb). The armament was to consist of two 23 mm cannon with a total ammunition capacity of 180 rounds.

Work on the project was completed in early 1947 and construction of the first prototype was undertaken at No. 115 Plant. On 17 June the prototype, designated Yak-23-1, was ready and immediately started factory trials with Victor V. Shelepchikov as the leading engineer. The first flight of the first prototype was made by M. I. Ivanov on 8 July 1947. The first flights displayed the huge excess thrust that resulted in high rate of climb and good manoeuvrability. During further trials a maximum speed of 932 km/h (578 mph) was achieved at low level and of Ma 0.845 at an altitude above 5,000 m. Many shortcomings

The first prototype of theYak-23. Photo was taken during manufacturer's flight tests.
Yakovlev OKB photo via author.

came to light during these trials, and these were gradually cured. Eventually factory trials were completed on 12 September and it was then that the Minister of Aircraft Production, Mikhail V. Khrunichev, asked the Commander-in-Chief of the Soviet Air Force Marshal Konstantin A. Vershinin to accept the Yak-23 for another stage of state trials. The second prototype, Yak-23-2, was earmarked for these, commencing on 11 November. G. A. Sedov was appointed the leading test pilot, and the other pilots of the flight test programme were P. M. Stefanovskiy, A. G. Kochetkov, A. G. Proshakov, Valentin I. Khomyakov,

Rear view of the first prototype Yak-23.
Yakovlev OKB photo via author.

One of the Yak-23s used by the Polish Air Force.
A. Gołąbek coll.

Yak-23 of OSL-5 Radom, Polish Air Force, early '50s.
Z. Karpiński coll.

Ivan M. Dzyuba, Leonid M. Kuvshinov and senior test pilot M. I. Ivanov. The trials were completed successfully in March 1948. The aircraft received very good reports in terms of manoeuvrability, and displayed easy take-off and landing (excess thrust, low weight, high lift). All pilots who flew the aircraft expressed favourable and positive opinions. Compared to the Su-9 and MiG-9 the aircraft had a much higher rate of climb. It displayed a clear excess thrust which made it highly manoeuvrable. There was also some criticism regarding longitudinal oscillations at speeds of Ma 0.865 and more. Another unwanted phenomenon was that the forward fuselage tended to lift by itself when speed was increased. Lack of air brakes was a clear shortcoming, making dog-fighting significantly more difficult. At an altitude of 5,000 m the time needed to reduce speed from 864 km/h (536 mph) to 637 km/h (395 mph) was 51 s. Lack of a pressure cockpit limited the aircraft's combat use at altitudes of 7,000-10,000 m (22,750 – 32,500 ft).

During this stage of trials test pilots conducted two practice air combats against the Tu-12 jet bomber. They developed the optimum method of attack from the rear, from a distance of 800 -1,000 m (2,600 – 3,250 ft), at an altitude of 8,000 -10,000 m (26,000 - 32,500 ft) at an angle of 35°-40°. Good results achieved in combat trials decided its usefulness for the air force. These and other advantages resulted in a positive overall opinion of the aircraft. Following completion of trials the Yak-23 was recommended for series production at No. 31 Plant. The document that outlined the role, the approval and acceptance of the type into service, was signed by Marshal K. A. Vershinin and by I. V. Markov, the Chief Engineer of the Soviet Air Force.

The second prototype Yak-23-2 (factory no. 115002, code "52") had by this time been lost. The crash took place on 14 July 1948 during a trial for the air display planned to be held the next day at Tushino. The aeroplane crashed

Yak-23 of OSL-5 at Radom (Polish Air Force Training Centre).
R. Karpiński coll.

near the main parade stand during a flypast of a group of fighter aircraft at an altitude of 100 m (325 ft). Bomber and transport aircraft flew overhead at the same time. A mysterious object hit the wing of the Yak-23, flown by M. I. Ivanov. Moments later a part of the wing was torn off and the aircraft started to fall, rotating about its axis. When it hit the ground the pilot was thrown out of the cockpit and killed immediately. It was later found that the Yak-23 was hit by a balance tab that detached from the tail of a Tu-14 bomber flying above.

Series production commenced in mid-1949. The delay resulted from lack of RD-500 engines, the first batches of which were delivered for La-15 fighters. The first production examples left the assembly lines of the Tbilisi plant in October 1949. From January to March 1950 one of the production examples was subjected to state trials at the GK NII VVS at Zhukovski. In the second quarter of the year military trials were conducted using 20 examples in an operational regiment near Grozniy. These trials revealed more shortcomings, such as failures of the cockpit hood jettisoning system, smoke in the cockpit

Yak-23 of Soviet Air Force.
Author's coll.

Two photos of UTI Yak-23 prototype in its original form.

Yakovlev OKB photo via author.

from the ventilation system, etc. But these minor faults did not affect the fate of the Yak-23.

The aircraft received seven-digit factory numbers, such as 3130725. A total of 313 were built, including 310 by No. 31 Plant in Georgia, and 3 Yak-23s (prototypes) by No. 115 Plant. No. 31 Plant manufactured twelve batches, of 25-26 examples each. Production was completed by the end of 1950 and its final portion was earmarked for export. In the Soviet Air Force Yak-23s were allocated to just a few air regiments of two military districts, Caucasus and Volga.

In parallel with the start of production of the Yak-23 work commenced on the training version, designated Yak-23UTI. The prototype UTI Yak-23-I was converted from no. 115001. The instructor's cockpit was located aft of the pupil's, at the very end of the fuselage, at the base of the fin. The prototype was subjected to testing during March-September 1949. Test results were not satisfactory and the prototype was converted to UTI-Yak-23-II form. In order to redesign the cockpit the fuselage was stretched by 200 mm. The instructor was seated in the front cockpit. A special periscope was fitted to allow the instructor a view of the rear cockpit. One 0.5 in. UB-12.7 machine gun was fitted for gunnery training. G. S. Klimushkin made the first flight, which resulted in further changes. The third prototype UTI Yak-23-III had the instructor's seat raised and the hood above it was bulged. The aircraft behaved well in the air but was not placed in mass production. This was due to the ending of Yak-23 series production and to rapid progress in development of the MiG-15UTI, which proved more promising.

In 1950 most Soviet fighter units converted to the latest swept-wing MiG-15 and La-15 fighters which exceeded the Yakovlev aircraft in capabilities and performance, so it was decided to deliver most Yak-23s to satellite countries. Yak-15s and Yak-17s had no export value due to their negligible combat capability. At the time Poland was the largest and most significant country of the Soviet block. The Korean conflict increased international tension globally. By then the combat value of the Yak-23 was rather limited, and it was significantly inferior to aircraft of the potential enemy, such as the F-86 Sabre then entering service in large numbers with US and other NATO air forces. Needless to say, Soviet block countries were given no choice and had to buy what they were given.

UTI Yak-23 with second version of the canopy at Chodynka Museum, Moscow.

M. Czaplicki

Yak-23.
1/72 scale.

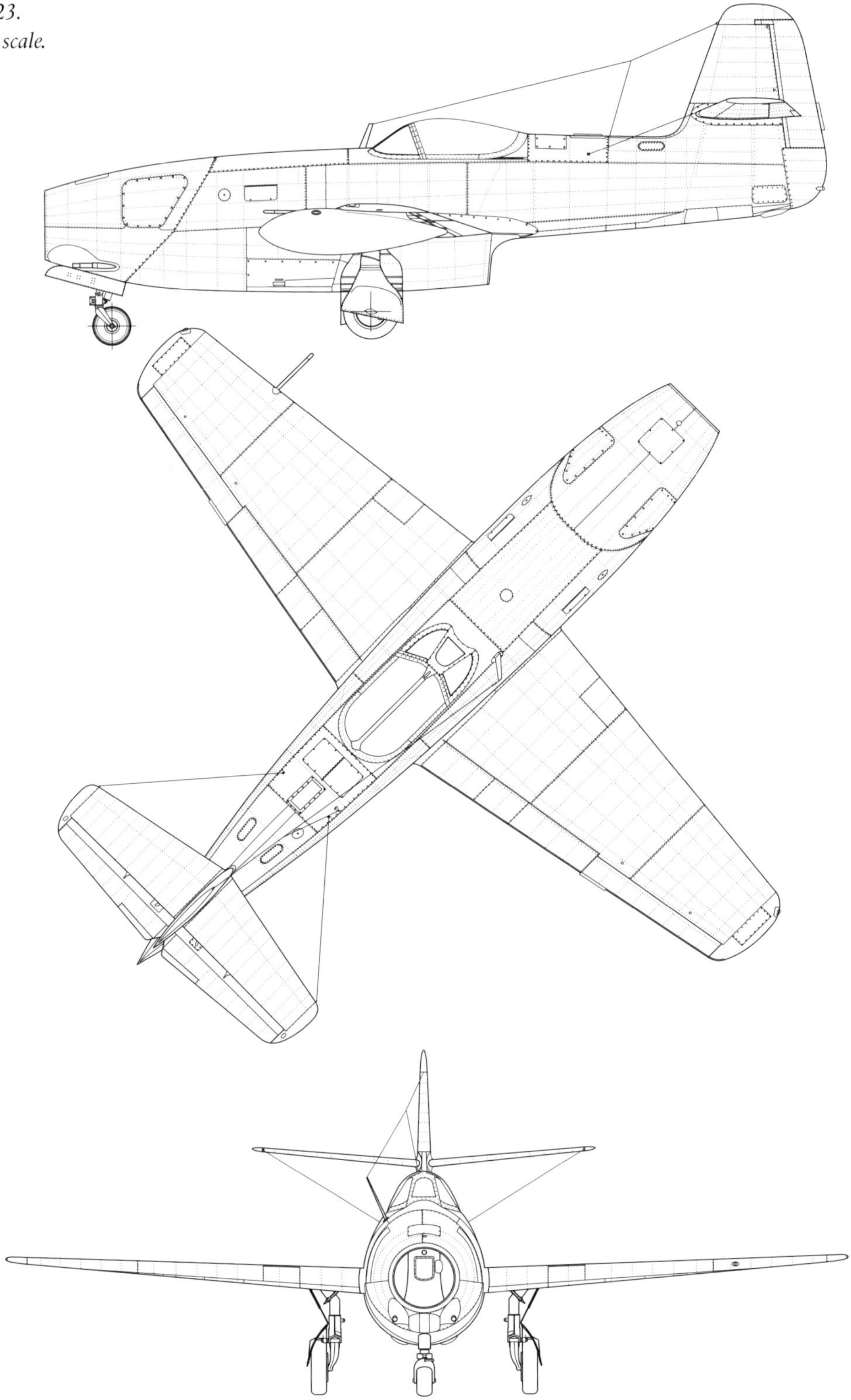

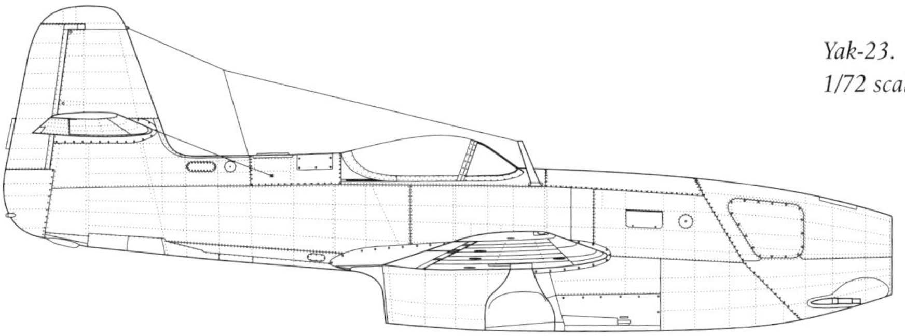

Yak-23.
1/72 scale.

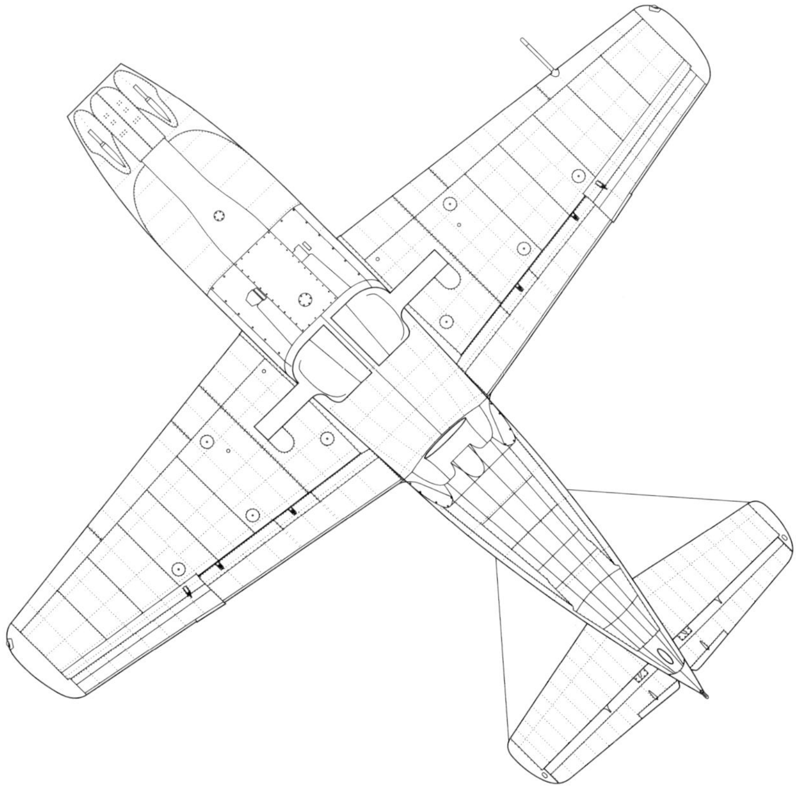

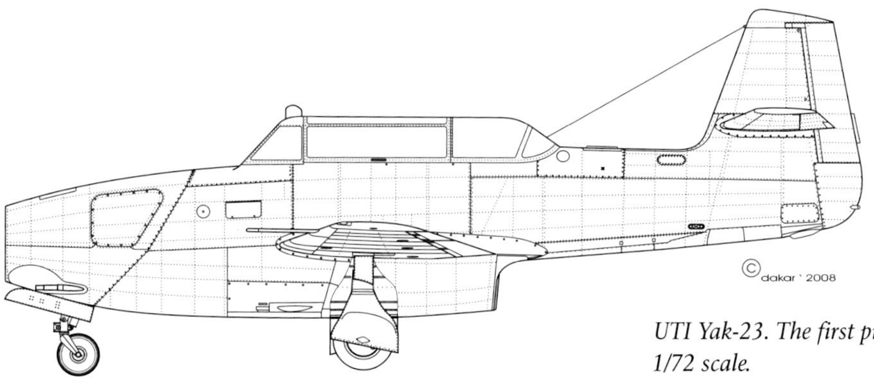

UTI Yak-23. The first prototype.
1/72 scale.

Yak-17 and Yak-23 exports

Poland

The late 1940s and early 1950s were a special period in international relations. The distance between the West and the East was growing. The Korean conflict put the world on the edge of another global war. Soviet leaders aimed at a radical increase in the military potential of the entire block. The Kremlin ordered individual countries to prepare plans for re-equipment of their armed forces. The scale of the process is best shown by the fact that the Polish air force was going to increase its potential fourfold, from seven to over thirty air regiments. This effort exceeded any real capabilities, so these plans were gradually altered. Apart from the reorganisation, equipment procurements had to be made, and production and training set up.

Reorganisation of the Polish air force commenced in the spring of 1950. According to earlier plans, jet-powered aircraft procurements had been planned as early as 1948, as the new class of equipment was entering Soviet Air Force service on a mass scale. That first plan had called for procurement of Yak-15s, but eventually it was decided to buy the newer Yak-17. Negotiations resulted in procurement of three Yak-17s, and licence production in Poland was con-

Yak-17UTI used by Polish Air Force at a static display during one of the military shows in the early '50s.
A. Tatarek coll.

Yak-17UTI used by Instytut Lotnictwa *(Aviation Institute, Warsaw) at Warsaw-Okęcie airfield.*
R. Gretzyngier coll.

sidered. At the same time, to allow conversion training of instructors three examples of the UTI Yak-17 trainer (also known as the Yak-17UTI, Yak-17V) were purchased. Three Yak-17s were shipped to Poland by rail. One aircraft was allocated to the 36th Special Air Regiment, while the other two were left in crates in a hangar at OSL-5 (*Oficerska Szkoła Lotnicza nr 5* - No. 5 Air Force College) Radom. One was going to be used as pattern for licence production. At the same time preparations commenced for conversion training of four pilots at the 159th Novorossiyskiy Guards Fighter Aviation Regiment of the Soviet Army Northern Group (based in Poland). Simultaneously the WSK factory at Mielec prepared to undertake licence production of the Yak-17 under the designation G-1. The first public display of the latest equipment was planned for the Aviation Day on 20 August. That day over 100,000 spectators watched a display by Lt. Col. Wasyl Gaszyn who performed aerobatics for several minutes, showing the manoeuvrability of the then mysterious Yak-17. In early 1951 the modern military air base at Warsaw-Bemowo was opened, and this became home to the newly-formed 5. DLM OPL (*5. Dywizja Lotnictwa Myśliwskiego Obrony*

Yak-23 of Polish Air Force at Tomaszów Mazowiecki airfield. Personal aircraft of por. pil. Jerzy Dębowski, sitting on the wing, summer 1956.
Author's coll.

Yak-23 of Polish Air Force. Air Show in Wrocław, September 1959.
W. Sankowski coll.

Przeciwlotniczej - 5th Fighter Aviation Division of the Air Defence), expanded from the 1. PLM (*1. Pułk Lotnictwa Myśliwskiego* - 1st Fighter Aviation Regiment). The new division was going to provide air defence of the Polish capital. In April three Yak-17UTI trainers arrived at Bemowo by rail from a Soviet unit. The aircraft went to the 1st PLM, which was going to become an unofficial jet aircraft conversion centre.

The outbreak of the war in Korea significantly accelerated the process of expansion and conversion of the Eastern block armed forces. Soon Soviet authorities decided to allocate larger numbers of new jet-powered aircraft for exports. Poland was granted the largest amount. As a result, on 6 January 1951 the first rail transport with crated Yak-23s arrived in Bemowo. They were

Yak-23s and its pilots of 40 Pułk Lotnictwa Myśliwskiego, *Polish Air Force (Air Regiment) during briefing just before a parade over Katowice in 1953.*
Author's coll.

Fitters during a break at Tomaszów Mazowiecki airfield, summer 1956.
F. Dziekan coll.

immediately assembled, flown and handed over to the 1. PLM. The aircraft were accompanied by four Soviet instructors, who started training instructor candidates. Meanwhile, it was decided to halt preparations for production of the Yak-17, this being replaced by the Yak-23 under the designation G-3.

Officially the Yak-23 was introduced into service with the Polish air force on 20 March 1951, when the first orders for training were delivered. At the time the regiment had 16 Yak-23s, 3 Yak-11 piston-engined trainers, 6 Yak-9P and 1 Yak-1M piston-engined fighters, and 3 Po-2 biplanes. The 2. PLM based in Cracow was the second Polish unit equipped with Yak-23 jets. The first three Yak-17UTIs for training were taken on charge in April 1951. The first flights by Cracow pilots took place during conversion training at Warsaw-Bemowo on 9 May 1951. Another Yak-17UTI was delivered in mid-May. The process of training continued as planned and by July it was planned to display the latest acquisition during a flypast. Yak-23 aircraft overflew the centre of Warsaw in a series of vics in line astern (five sections of three Yak-23s each). The newly formed 13. PLM of the 5. DLM OPL was the next unit to become equipped with the Yak-23. On 13 August the unit took delivery of 13 aircraft from the 7th, 8th, 9th and 10th production batches. Another set of 7 Yak-23s went to

Yak-23 of Polish Air Force and its fitters.
F. Dziekan coll.

Yak-23 of Polish Air Force in winter, late '50s.
Author's coll.

the 13. PLM a month later. These aircraft were handed over by the 1. PLM, then converting onto the latest MiG-15s. Meanwhile it was decided to re-equip the entire Warsaw-based 5. DLM OPL with MiG-15s, while the Cracow-based 7. DLM OPL took over all 64 of the Yak-23s delivered up to then. These Yak-23s equipped three regiments, the 2., 39. and 40. PLM of the division. In December

Yak-23 at Radom airfield.
Author's coll.

another batch of 3 Yak-17UTIs was purchased and transferred to the forming 11. PLM at Poznań-Krzesiny. Meanwhile, with the advent of 1952 new divisions started to be formed: 6. DLM OPL at Wrocław, 9. DLM at Malbork, 10. DLM OPL at Słupsk and 11. DLM at Świdwin. In the spring of 1952 the 29. PLM of the 9. DLM was formed at Warsaw-Bemowo and equipped with 25 Yak-23s, this being the fourth regiment at the time equipped with Yak-23s and Yak-17UTIs. Following its transfer to Malbork, 11 of its Yak-23s were handed over to the 26. PLM. Thus by mid-1952 all Yak-23s delivered to Poland were in five units:

- 2. PLM – Kraków-Balice: 24+2 Yak-23;
- 39. PLM – Mierzęcice Zawierciańskie: 19 Yak-23;
- 40. PLM – Mierzęcice Zawierciańskie: 19 Yak-23;
- 26. PLM – Malbork-Królewo: 11 Yak-23;
- 29. PLM – Malbork-Królewo: 14 Yak-23.

The next flypast in Warsaw with Yak-23s participating took place on 22 July 1952. These were flown by 39. PLM from Mierzęcice. In the autumn of 1952 another newly formed division, the 10. DLM, started to be equipped with jets (25. and 28. PLM). In the absence of new Yak-23 deliveries the 25. PLM was equipped by reducing the numbers of aircraft in other units. The 25. PLM started to be formed from the second flight of the Cracow-based 2. PLM. The regiment received 10 Yak-23s from four other units. Early 1953 brought reorganisation of the WL i OPL OK (*Wojska Lotnicze i Obrony Przeciwlotniczej Obszaru Kraju* - Aviation and Country Area Anti-Aircraft Defence Forces). It was decided at the time to re-equip first-line units with the latest MiG-15s, while second-line units received Yak-23s. Therefore Yak-23s were transferred from the 29. and 25. PLM to the 6. DLM (3. and 11. PLM).

In the early 1950s Poland and other Soviet satellite countries interested Western intelligence. One of the methods of penetrating the airspace of the

Yak-23 of OSL-5, Polish Air Force.

P. Słoma.

Por. pil. Stanisław Kopacz in his Yak-23 at Tomaszów Mazowiecki airfield.
S. Kopacz coll.

Eastern block was to use various types of balloons. Some were typical reconnaissance units, some served for propaganda, and others for meteorological duties. The expanding Polish Air Force had an observation-reporting system, radar posts and jet fighter aircraft. During that period such 'unwelcome guests' were intercepted by most air units. One such episode was described by Capt. Józef Mizera from the 29. PLM based at Orneta.

"*On a cool September morning together with my colleague Masternak we checked a pair of previously prepared Yak-23s that stood in the area of the readiness pair. As soon as the technicians had left the alert sounded: to the aircraft! Having performed all the necessary actions we took off. A moment later the ground controller ordered my wingman to turn back. So I was going to continue my task alone. Another instruction from the ground specified the heading and altitude, and*

Yak-23 of OSL-5 Radom.
P. Słoma.

I followed it immediately. Having broken through clouds and reached an altitude of 7,200 m I asked about the target of the interception. I was told to continue climbing to 8,000 m. Slowly I started to feel the cold and the characteristic sounds of the pressure system working. After a few moments the ground controller told me that a slow moving target was 30 km ahead of me. After a few corrections of heading I was ordered to prepare my armament. At this time the ground controller told me that I was 5 km from my target, but I still could not see it. After another manoeuvre a huge container appeared suddenly in front of me, I realised it was a balloon with a suspended pod. I passed it several hundred metres to one side, unable to open fire. I made another attack, and this time the unusual target fooled my intuition. The balloon was large, some 25 m in diameter. Moreover, its speed was very low, it seemed to be standing still. I made another manoeuvre to attack, this time aiming precisely and fired a burst from some 1,000 m, a few seconds later I saw the rounds miss the target. I was angry that such a passive target was impossible to bite. I came in for another attack, this time calculating deflection for the distance and ballistic path of the rounds, I aimed at the upper portion of the envelope. Another series seemed too high, but ballistics worked and the rounds went into the middle of the balloon envelope. A moment later I saw a flash and the envelope was torn apart. Then the balloon started travelling rapidly towards the ground. This was my first combat sortie, and I gained some experience against unusual air targets".

In the first half of 1953 four more Yak-17UTIs were delivered to Poland, going to the 2., 3. and 11. PLM. In mid-1953 Yak-23s were on charge with six regiments: 2., 3., 11., 26., 39. and 40. PLM, and in HQ sections of the

One of the pilots of 2 PLM in Kraków, Polish Air Force. Yak-23 serial 1122.
R. Gretzyngier coll.

3. KLM (*3. Korpus Lotnictwa Myśliwskiego* - 3rd Fighter Aviation Corps) and 11. DLM.

As mass licence production commenced of the MiG-15, under the designation Lim-1, Yak-23s were gradually phased out of first-line units. At first Yak-23s were transferred from the 40. PLM to regiments of the 6. DLM. In the spring a batch of Yak-23s was delivered to the OSL-5 at Radom, where in August cadet officers commenced training on jet-powered aircraft. Then in the autumn of 1953, as the 11. PLM converted onto MiG-15s, some of its Yak-23s were transferred to the 21. PLZ (*21. Pułk Lotnictwa Zwiadowczego* - 21st Scout Aviation Regiment) in Poznań and to the second flight of the OSL-5 based at Tomaszów Mazowiecki. At the end of August 1954 the last Yak-23s from the 2. PLM were handed over to the informal maintenance base of these machines at the OSL-5 Radom, then being established. Eventually all serviceable Yak-23s were allocated to the OSL-5, a few dozen of them assembled at Radom. Some were unserviceable, and were cannibalised for spares. Two examples were handed over to the TSWL (*Techniczna Szkoła Wojsk Lotniczych* - Air Force Technical School) at Zamość, where they were used for ground instruction. By May 1954 the 21. PLZ, now moved from Poznań to Sochaczew, remained the sole combat unit to use Yak-23s. Soon afterwards the 21. PLR converted onto Lim-1s and -2s, and from then on the OSL-5 was the sole operator of the Yak-17UTIs and Yak-23s.

Between 1953 and 1958 two training flights of the OSL-5, 2. EPB (*2. Eskadra Pilotażu Bojowego* - 2nd Combat Flying Flight) and 3. EPB at Tomaszów Mazowiecki used several dozen Yak-23s and nine Yak-17UTIs. Following conversion training of selected instructors onto Yak-23s, training of officer cadets commenced. Specific flying characteristics of the Yak-23 were noted by the cadet officers. Jan Paweł Graczyk, then under training with the 2. EPB, recalled:

"*We made our flights with instructors on MiG-15UTIs, as the flight had no Yak-17UTIs left on charge. There were mostly exercises at elementary level, circuits, to the zone for elementary flying, formation and cross country flying. Then, on the Yak-23 we first went through the theory in the cockpit, then taxiing and acceleration, known colloquially as 'run-ups'. It was this element that gave cadet officers the most emotions and problems. Many times such 'run-ups' ended in lift off and an unexpected circuit followed by a landing. Such unexpected manoeuvres*

The last days of Yak-23 used by Instytut Lotnictwa in Warsaw.
R. Gretzyngier coll.

Yak-23 with civil registration SP-GLK (probably the only civil Yak-23 in the world) used by Instytut Lotnictwa in Warsaw, late 50s.

R. Gretzyngier coll.

were usually punished with one week of confinement to quarters. The Yak-23 was a fascinating type for us cadet officers, easy to fly. It had its kick, lifting off the runway relatively quickly, usually in the middle of the runway, and you could have 800 m altitude at the first turn. It was also easy to land, but it did not tolerate more serious errors such as bounces or nose wheel touchdown".

Yak-17UTIs were successively withdrawn from units starting from 1952 (no. 3120008 from the 39. PLM on 26 May). The last two aircraft, nos. 3120127 and 3120037 from the OSL-5, were withdrawn on 7 October 1955. In early 1957 the air force transferred two Yak-17UTIs to the civilian ILot (Instytut Lotnictwa - Aviation Institute), one of these was registered SP-GLM while the other was used as a source of spares. In the first half of 1957 the OSL-5 still had 21 Yak-23s. As more MiG-15s and Lim-1s arrived at the Radom establishment, the ageing Yak-23s were withdrawn gradually. In early 1959 an informal depot of the withdrawn Yak-23s was established at Radom. The wrecks held there provided spares for the few Yak-23s still used by the 60. and 61. LPSz-B (*Lotniczy Pułk Szkolno-Bojowy* - Training-Combat Air Regiment). On 1 September 1959 the last 32 Yak-23s were struck off charge by the Polish air force. At the same time three Yak-23s were displayed as memorials, one in Kraków-Bielany, one in Zakopane, and one at the Częstochowa flying club. This ended the eight year period of Yak-23 operation in the Polish military aviation.

In 1956 two Yak-23s were handed over to the *ILot*, registered SP-GLK and SP-GLL. Only one example, SP-GLK (factory no. 3130925) was used for flying trials. On 21 September 1957 ILot test pilot Andrzej Abłamowicz used this aircraft to establish two world rate of climb records: 119s to 3,000 m (25.2 m/s) and 197s to 6,000 m (30.5 m/s). The aircraft continued to fly with ILot until 28 November 1961, when it was struck off charge.

During eight years of operation of the Yak-23 by the Polish military aviation, the design proved safe and forgiving to human error. There were 7 tragic crashes for the 101 aircraft in use, a much lower rate than for other types of jet-powered aircraft. In most cases of malfunction or damage the pilots escaped without serious harm. The largest number of Yak-23 accidents was noted in the 11. PLM – six, including one fatal, two crashes without casualties and three aircraft damaged. Most accidents in the entire air force took place in 1953: fourteen (5 fatal crashes, 4 crashes without casualties and 6 aircraft damaged).

Czechoslovakia

Czechoslovakia was the second principal satellite country in the Soviet block, and it also underwent a five-year armed forces expansion plan during the late 1940s/early 1950s. The Czechoslovak air force had a different situation, as at the end of the war they captured German Me 262 Schwalbe jets assembled in Czech territory, and used them under the designation S-92. They had a total of 9 serviceable S-92 fighters and 3 CS-92 two-seat combat trainers, grouped in one unit.

Almost at the same time as Poland, Czechoslovak authorities obtained permission to purchase a licence for Yak-17 production. To this end in the first half of 1950 the Mlada Boleslav aircraft factory near Prague received one example, to be used as pattern for production. The Yak-17 was going to be built under the designation S-100. However, the changing political-military situation resulted in changes to these plans. The sole Yak-17 never flew, and after its licence production was abandoned it was handed over to an aviation museum. The Yak-23 was the next aircraft purchased in the USSR with licence production in mind. In November 1950 the necessary documents were signed for licence production of the Yak-23 under the designation S-101, and of the RD-500 engines as the M-02 by the Motorlet plant. In early 1950 the aircraft, together with the MiG-15, was shown by one of the Soviet units to a Czechoslovak delegation with the President, Klement Gottwald. According to plans a formation of six S-92s and eleven S-101s were going to perform a flypast on the Republic Day in April. The first 12 Yak-23s arrived in Czechoslovakia in late 1950/early 1951. According to some sources, on 1 April 1951 the Czechoslovak Air Force had 10 Yak-23s, grouped in the 5th Air Flight at Cheb. Official presentation of the new equipment was planned for 6 May 1951, the anniversary of liberation of Czechoslovakia. Nine Yak-23s were going to be the principal component of the military parade. Probably during that period Czechoslovakia received another nine Yak-23s (unconfirmed data). During that period the first MiG-15UTIs and MIG-15s arrived at Blade air base, these being newer aircraft with much better performance than Yak-23s. In this situation training on Yak-23s became a secondary matter. Meanwhile it was decided to form the 3rd Fighter Division,

Yak-17 (S-100) at Mlada Bolesav factory. The only Yak-17 in Czechoslovak markings.

Stratus coll.

A Yak-23 (S-101), coded KR-17, one of only eleven aircraft of this type delivered to Czechoslovak Air Force in late 1950, belonged either to the 5th Fighter Squadron or to the VSL Air Force Jet Training Centre at Milovice-Mladá, as both units used the same 'KR' code to mark their aircraft.

composed of the 1st, 5th and 6th Fighter Regiments. At the same time the 5th Fighter Division started to be formed, based on the 11th Fighter Regiment. During June-August 1951 selected personnel of the 11th Rgt converted onto Yak-23s. In mid-December 1951 the 11th Rgt moved from Blade to Zatec. In early 1952 the 11th Rgt had 19 pilots, and 11 Yak-23s (S-101s) and 5 MiG-15s. From the spring 1952 pairs of Yak-23s were held at readiness at Cheb (2 aircraft) and Zatec (4 Yak-23s). On 16 October one Yak-23 (factory no. 310103, code KR-12) was lost in accident with its pilot, Lt. Jizi Jermak.

In 1953 new units were formed. The 51st Air Regiment was expanded into the 22nd Air Division, and in May the 166th Fighter Division. At the time the 1st flight of the 51st Rgt received all the Yak-23s, moved from Zatec in the spring 1953. According to unconfirmed information, in the spring of 1953

***Below:** This Yak-23 (S-101) Flora, HX-51 (s/n 10101), was photographed outside the hangar at Prague-Kbely airfield. It carries the code of the 3rd Squadron, 3rd Air Regiment, which had it on its strength in the period 1954-56 and then based at Brno. In 1969 it was restored by VLO Air Force Repair Works and is currently preserved at the Aviation Museum in Prague-Kbely. Both photos courtesy 4+ Publishing Co.*

An in-flight shot of a trio of Czechoslovak Yak-23s (S-101), coded IF-15, IF-17 and IF-13. The aeroplanes belonged to the 1st Squadron of the 51st Air Regiment based at Milovice-Mladá. The nose codes and serial numbers (painted under the tailplane) were black, while the national insignia were outlined in blue.

Collection 4+ Publishing Co.

Poland purchased 10 Yak-23s from Czechoslovakia and allocated them to the OSL-5 Radom. On 1 October some unit names were changed, and the 51st Rgt was renamed the 7th Rgt, while the 166th Division became the 2nd Fighter Division. In 1954 the 7th Rgt moved to Kosice, at the same time all remaining Yak-23s were allocated to the 3rd flight of the 3rd Rgt at Brno. In early 1955 the regiment had 12 Yak-23s, including 11 serviceable ones.

In 1956 it was decided to withdraw Yak-23s from the Czechoslovak air force. At the time information appeared that 7 serviceable aircraft were sold to Bulgaria. This was the end of Yak-23 in the Czechoslovak air force. The number of Yak-23s in use cannot be verified due to lack of complete documentation, and to contradicting eyewitness accounts. Nevertheless it is probable that 21 Yak-23s were used, of which 10 were sold to Poland in 1953, 7 to Bulgaria, 1 Yak-23 crashed and 1 was handed over to a museum.

Bulgaria

Bulgaria was the third country to purchase Yak-23s from the USSR. The first 12 Yak-23s arrived in Bulgaria in early 1951. The aircraft were allocated to the 19th Fighter Regiment formed in March 1951 at Graf Ignatievo. Not long before that Bulgaria took delivery of several Yak-17UTIs that were used for conversion training of instructors on jet-powered aircraft. On 1 June 1951 Maj. Vasil Velichkov became the first pilot to fly a jet-powered aircraft. During take-off the engine stopped, but the pilot made a safe forced landing in a field outside the

airfield. The flypast over Sofia on 1 May 1952 included a formation of 9 Yak-23s and 9 MiG-15s. Yak-23s were used in other Bulgarian units, including the 2nd Training-Combat Air Regiment and the Georgi Benkovski Flying School. Yak-23 readiness pairs were often scrambled to intercept violators of Bulgarian air space. During 1956 (date unknown) pilot Gradinarov managed to down a reconnaissance balloon at 39,130 ft. by accidentally catching it with his wing. In 1952 a decision was made to rearm the 4th Fighter division (consisting of 11th, 14th and 18th regiments) with jet fighters. Because of this, more Yaks were bought from the Soviet Union. The same year Soviet pilots brought 72 more planes, divided between the 11th and 18th regiments.

Bulgarian Yak-23 during preparation for the next flight.

Yak-23 during night exercises.

All Stefan Boshniakov coll.

Two photos of Bulgarian Yak-23 during overhaul.
Stefan Boshniakov coll.

In early 1956 Czechoslovak authorities sold 7 Yak-23s to Bulgaria for 50,000 crowns, and a batch of engines for 25,000 crowns. During 1954 10 Yak-23s were sent to Air Force School "Georgi Benkovski" at Dolna Mitropolia, and served there till 1957.

The last regiment that was armed with Yak-23s was the 20th fighter-bomber regiment (20th IShAP) at Kroumovo airbase. The last planes were retired in 1959.

Two photos of Yak-23 just after the flight.
Both Stefan Boshniakov coll.

Romania

Romania was another user of the Yak-23 and Yak-17UTI. As with other communist countries, Romanian authorities prepared a plan of accelerated expansion and modernisation of their armed forces in connection with the Korean war. The plans for the air force included expansion of structures and procurement of jet-powered aircraft. The first group of pilots and ground crew underwent conversion training on jet Yaks in the USSR. The first flight in a Yak-17UTI was made by Maj. Aurel Raican on 26 June 1951. The numbers of Yak-17UTIs or Yak-23s acquired are unknown. According to unofficial sources the first 12 Yak-23s reached Romania in early 1951. Altogether the Romanian air force used some 60 Yak-23s, allocated to regiments of the 97th Fighter Division. In 1953 Romania took delivery of about 60 MiG-15s, which entered service with three fighter regiments. Rapidly changing situations and introduction of the MiG-15 into service resulted in Romanian procurement of only a small number of Yak-23s. Romanian machines also participated in interceptions of air space violators. The first successful interception was performed by Maj. Dymitr Bajaur. In 1956 Romanian designers of the ARMV-2 (*Atelierele de Reparatii Material Volant*) at Bucharest-Pipera developed a prototype of a two-seater training version. The aircraft was fitted with an instructor's cockpit and was designated Yak-23 DC (*Doubla Comanda)*. The training version never entered series production.

Romanian Yak-23s are connected with an interesting story that remained secret for over 40 years. On 24 July 1953 a Yak-23 flown by Lt. Michail Dijakoanu left a Romanian airfield, and headed for Yugoslavia, landing soon after. Since 1948 the communist authorities of Yugoslavia had enjoyed political

Romanian Yak-17UTI.
D. Bernad coll.

***Above:** Romanian Yak-23 of unknown unit.*

***Middle & bottom:** Two photos of crashed Romanian Yak-23.*

All D. Bernad coll.

independence from the USSR. At the time President Josif Broz Tito took up a US offer and acquired some American armament, including F-84 Thunderjet jet fighters. Yugoslav authorities were grateful to the Americans for their help. The pro-Western policy of Yugoslavia gave Western intelligence services more opportunities. In August 1953 the news of the defection of a Romanian pilot in a Yak-23 to Yugoslavia reached the CIA resident in Belgrade. Negotiations started in this significant but delicate question: the Americans wanted to borrow the Yak-23 and ferry it by air to the USA for testing. By early October the secret agreement was ready and teams prepared for the secret mission, codenamed "Zeta" by the Americans. All national markings and codes were removed from the Romanian Yak-23. Aircraft engineering experts dismantled the aircraft and put it in crates that matched dimensions provided by the Americans. In early October 1953 five representatives of the Yugoslav air force, Col. Kosta Sivčev (designer), Lt. Col. Milorad Spasič, Branislav Jovanović from the Vazduhoplovni Tehnički Institut at Žarkov, Maj. Miroslav Boras (aircraft armament expert), and Maj. Zvonimir Kos. arrived at the agreed location on one of the airfields with the special load prepared. According to plan a Douglas C-124 Globemaster II of the 4th Transport Squadron 62nd Transport Wing arrived to collect the mysterious cargo. The crated Yak-23 was accompanied by the Yugoslav team. The aircraft was delivered to Wright Patterson AFB at Dayton, Ohio. There it was collected by a technology intelligence group for static trials and flight testing. The aircraft was uncrated and assembled, given a military number (FU-599) and US national markings. The Yak-23 was located in a special zone of no. 145 hangar. Tests revealed that the machine's tyres were excessively worn, so replacements were quickly obtained from Goodyear. Soon trial engine runs were undertaken, conducted early in the morning at the outskirts of the airfield. Prior to the first test the start-up pump failed, and was replaced from an original Derwent V engine. Flying trials were performed by test pilot Tom Collins. The first flight took place on 4 November and after eight flying tests, on 25 November the aircraft was dismantled and crated. It was loaded into a C-124 and flown with three mysterious guardians to Pancevo in Yugoslavia. This unusual action by two intelligence services ended successfully. After six months the thoroughly tested Yak-23 was returned to the Romanians, who never guessed how far the aircraft had travelled.

Romanian Yak-23 with national markings removed, just before tests in the USA.

Author's coll.

Three more photos of the Yak-23 during tests in the USA.

Author's coll.

People's Republic of China

In late 1950/early 1951 the USSR delivered a large shipment of armaments to China, including aircraft of various types. Most of these were piston-engined aircraft, such as the Il-10, La-11, Tu-2. Rapid escalation of hostilities in Korea forced the USSR to give more assistance to the Chinese army. The Soviet authorities decided to deliver jet-powered aircraft – Yak-17UTI trainers and MiG-9 and MiG-15 fighters. Several groups of Chinese pilots were trained by instructors from the USSR in schools at Peking and Hang-Zhou, and also at their own 29th Guards Fighter Aviation Regiment. With the radically changing situation at the Korean front MiG-15s arrived in large numbers, constituting real opposition to US F-86s. The Yaks were used in small numbers in training units.

Hungary

No credible and documented information about Yak-23 and Yak-17 operations by the Hungarian air force could be found.

Altogether in the early 1950s the USSR exported to its satellite countries some 200-210 Yak-23s (Poland - 90, Czechoslovakia - 21, Bulgaria - 40, Romania - 60, Hungary - 12), 20-30 Yak-17UTIs and a few Yak-17s (including 3 to Poland and 1 to Czechoslovakia).

Yak-23 structure

The aircraft was a lightweight, all-metal fighter for low and medium altitude operations. It was built in a conventional layout of a mid-wing monoplane with unswept wing and tail surfaces. The fuselage was in a stepped layout, with stressed skin semi-monococque structure. The forward fuselage could be detached for easier maintenance of the power plant. Outer fuselage skin was made of 1 mm D16AWTL duralumin sheets, the inner skin was made of 0.8 mm sheets. Central air intake was located at the front, split into two symmetrical ducts to pass the cockpit. Nose wheel well and two 23 mm cannon bays were located in the bottom forward fuselage. The rear fuselage was made of 27 frames, 4 longerons and a number of stringers. The forward, reinforced, frame no. 1 was the base for power plant attachment. The main fuel tank was located further aft. The cockpit was located in the central fuselage, with the ejection seat supported on frame no. 14. A special cover was fitted in the lower central section, near the exhaust nozzle, to protect the main wheel wells located underneath. The underside of the rear fuselage was covered with heat resistant steel to protect the structure from thermal damage.

The main plane consisted of two unswept tapered wings with varied airfoil: laminar flow S-1-12 from the fuselage to rib no. 6, blending into KV-Z-12 further out. The two-spar wing had 3.5° dihedral. Each wing was made of 17 ribs and was covered with 1.2-1.8 mm duralumin panels. The wings were fitted with ailerons and flaps along the trailing edge. Frise type ailerons were mass- and aerodynamically-balanced. Slotted flaps of 1.603 m^2 area were attached at three points. Pneumatically operated, they deflected 25° for take-off and 60° for landing.

The conventional tail had a tapered planform. The fin of 1.0955 m^2 area was built as one section with the rear fuselage. The rudder deflected symmetrically

Polish Yak-23 after flight. **Author's coll.**

30° to each side. Single-piece tailplane had 5° dihedral. The two-piece elevator, fitted with balance tabs, deflected 33° upwards and 25° down.

The aircraft was powered by a single RD-500 turbojet engine with maximum static thrust of 1,560 daN (1,590 kG/3,500 lb), fitted with a single centrifugal compressor and nine cylindrical combustion chambers. A single-stage gas turbine and exhaust nozzle were fitted aft of the chambers. The engine was mounted diagonally at 4° 30´ to the aircraft centre line. Empty weight of the engine was 581 kg, its length was 2,060 mm and diameter 1,090 mm. The engine was fed with jet fuel (kerosene), held in five fuselage-mounted tanks with internal capacity of 910 l and two 190 l wing tip tanks (total capacity 1,290 l).

The undercarriage was tricycle, with the nose wheel (400x150 mm) retracting forward. Main wheels (570x140 mm) retracted into the fuselage sides, whilst their struts were lying horizontally inside the wings. Oleo shock absorbers with double rebound system were fitted. Nose wheel leg pivoted 55° to port and to starboard. Main wheels were fitted with drum brakes. There were two systems of undercarriage position indicators, electric in the cockpit and on the legs, as well as mechanical ones in the form of rods protruding from wing upper surface.

The cockpit was located in the central section of the fuselage, with a fixed windscreen fitted with an armoured glass panel and a rearward sliding hood. A lightweight structure ejection seat with a seat-type parachute was fitted. The ejector explosive charge was located in the rear part of the seat, whilst the ejector actuation mechanism was in the starboard armrest of the seat. The seat and main controls (stick and rudder pedals) were mounted on the floor that was supported by the wing structure. The instrument panel was ahead of the pilot, with the ASP-3M semi-automatic gyro gun sight above it. The instrument panel was in three sections, with 12 main flying instruments in the centre, including the M-46 Mach meter, engine control indicator, PDK-45 compass, AGK-47A artificial horizon, VD-15 altimeter, KUS-1500 airspeed indicator, VAR-75 rate of climb indicator, exhaust gas temperature indicator, fuel pressure indicator and heading indicator. Auxiliary controls, the undercarriage lever, etc. were located

Front view of Polish Yak-23.
W. Hołyś.

on both sides of the panel. The cockpit was fitted with an oxygen system of 8 litres capacity with the KM-16 mask. Heating and ventilation of the cockpit was provided by a smoothly adjusted system fed with warm air bled off from aft of the engine compressor. Air for ventilation was fed from the atmosphere via an intake in the engine feed duct.

Armament consisted of two 23 mm NR-23 cannon located in the lower forward fuselage, with 90 rounds of ammunition per gun. An ASP-3N gun sight was used for aiming. Gunnery control was provided by the S-13 camera gun fitted in upper forward fuselage above the air intake. The aircraft could take two bombs of up to 60 kg, attached in place of external fuel tanks.

The electric system of the aircraft was fed from the 1.5 kW GSK-1500 generator and 12A-10 battery. Radio and electronic equipment included the RSI-6K radio, RPKO-10M radio-direction finder, and SCh-3M IFF.

Specification

	Yak-17	UTI Yak-17	Yak-23	UTI Yak-23
Length	8,700 (mm) 28.54 (ft)	8,700 (mm) 28.54 (ft)	8,120 (mm) 26.64 (ft)	8,320 (mm) 27.30 (ft)
Span	9,200 (mm) 30.18 (ft)	9,200 (mm) 30.18 (ft)	8,730 (mm) 28.64 (ft)	8,730 (mm) 28.64 (ft)
Height			3,300 (mm) 10.83 (ft)	3,300 (mm) 10.83 (ft)
Wing area	14.85 (m^2) 159.80 (sq ft)	14.85 (m^2) 159.80 (sq ft)	13.50 (m^2) 145.3 (sq ft)	13.70 (m^2) 147.5 (sq ft)
Empty weight	1,852 (kg) 4,083 (lb)	2,084 (kg) 4,594 (lb)	1,980 (kg) 4,365 (lb)	
Take-off weight	2,890 (kg) 6,371 (lb)	2,906 (kg) 6,407 (lb)	3,036 (kg) 6,693 (lb)	3,410 (kg) 7,518 (lb)
Maximum speed (kph/mph) - near the ground - at an altitude of 5,000 m - at an altitude of 10,000 m	 700/435 786/435 	 702/436 748/465 	925/575 910/565 868/539	 816/507 850/528
Rate of climb	12 (m/s) 40 (ft/s)	11 (m/s) 36 (ft/s)	47 (m/s) 154 (ft/s)	
Service ceiling	12,750 (m) 41,800 (ft)	12,100 (m) 39,700 (ft)	10,000 (m) 32,800 (ft)	
Range	395/717 (km) 245/446 (mi)	370 (km) 230 (mi)	1,030 (km) 640 (mi)	
Take-off run	635/720 (m) 695/787 (yd)	640 (m) 700 (yd)	550 (m) 602 (yd)	
Landing run	650 (m) 711 (yd)	695 (m) 760 (yd)	650 (m) 711 (yd)	
Engine type	RD-10	RD-10	RD-500	RD-500
Engine thrust (kN)	9.8	9.8	15.9	15.9
Gun armament number/calibre/type	2x23 mm NS-23 cannon	none	2x23 mm NR-23 cannon	1x0.5 in. UB-12,7 machine gun
Number of rounds	2x60	none	2x90	

Short evaluation of the Yak-23

The design was evaluated many times, both by special committees during acceptance trials, and by groups of experts outside the USSR (including the technical intelligence group at AFB Wright Patterson). The aircraft and its operation were also evaluated by many military pilots who were able to get to know the design in detail. During design work some limitations were assumed, and these significantly affected its performance and capabilities. The design had a low empty weight (1,980 kg), achieved by removing various systems. Its aerodynamic layout gave it excellent manoeuvrability, whilst huge excess thrust gave it a relatively high rate of climb (up to 47 m/s [153 ft/s]). Its conventional layout ensured ease of flying, while take-off and landing runs within 500-600 m allowed it to be based on airfields with short runways.

These features are illustrated by the following accounts from the period when the type was used in the Polish air force.

Col. Stanisław Wdowczyk describes his personal experiences while serving with the 2. and 39. PLM (7. DLM OPL) at Cracow and Mierzęcice.

"TheYak-23 was an extraordinary machine, as the pilots used to say it had a soul. Thanks to its design features and excess thrust engine the aircraft was a reliable machine in the hands of a pilot. Lt. Józef Groszewski learned that when, while coming in to land as a wingman, he entered the slipstream of the leading aircraft. In just one moment his aircraft was turned upside down. The pilot did not lose control, but opened up, and hanging upside down in his harness pushed the stick. The aircraft started to gain height, and having reached safe altitude the pilot resumed the correct position."

Por. pil. Zdzisław Czech of the Polish Air Force exiting his Yak-23, Mierzęcice airfield, early '50s. **Author's coll.**

"Lt. W. Jędrych made an error while in level flight at an altitude of 500 m. As a result his aircraft started to spin. His subsequent actions were correct, according to well trained habits. He immediately did the right things and recovered the aircraft from the spin. To his surprise the pilot realised that his fine dive ended at a very low altitude. He was just 50 m away from a crash."

Capt. Henryk Rzemieniecki had an adventure that proved the high strength of the airframe:

"While coming in to land at the airfield in Cracow, at an altitude of some 30 m the pilot realised all his controls were locked. Probably the ground lock was not secured properly before take-off and it switched on by itself. The uncontrolled aircraft hit the concrete runway at a shallow angle. It made two jumps, fortunately without banking to one side. During this time the pilot managed to unlock the controls and ended his

landing run safely. In this case the Yak-23's particular feature paid off: stability which, combined with structural strength and correct actions by the pilot, meant that the event ended safely."

Lt. J. Żarnowski's adventure:

"*While carrying out his duty in the air, his engine stopped through lack of fuel. The pilot decided to make an emergency landing in a field. Unfortunately there was no glade nearby, just more than enough forest area below. There was no other choice – he had to try this unfriendly woodland. Moments later the good old Yak-23 flew into the wood, losing its wings. Several dozen meters further the fuselage came to a stop. There was silence. The most interesting thing was what the rescue team found when they arrived there. Several dozen meters away from the badly damaged aircraft, the pilot was lying under a tree, his head resting on a bag with the unused parachute in it. When the doctor came running to the pilot and asked how he felt, the shocked pilot said 'I am broken throughout and dying'. 'But who has put the parachute under your head?' The pilot replied 'I first escaped from the cockpit, but then returned to pick up the parachute, otherwise it might burn'.*"

This happy ending of the crash landing in a wood was largely thanks to the strong structure of the Yak-23. The forward and angled location of the engine made it absorb most of the energy during collisions, protecting the cockpit above it.

Lt. Stanisław Niedźwiecki went through an unusual accident.

"*Having carried out his task, Niedźwiecki's aircraft was coming in to land at Cracow airfield from the direction of Nowa Huta district, then under construction. Suddenly his lowered undercarriage hit the roof of a wooden shed. Construction workers had erected it exactly along the runway, just outside the airfield. The shed did not survive the collision, but the aircraft escaped safely, with only minor damage. In this case, again, the happy ending was thanks to the excellent longitudinal stability of the Yak-23.*"

In this short evaluation it is worth mentioning experiments and trials in which this aircraft was used to develop operational tactics in the conditions at the time. As mentioned before, during the first stage of introduction of jet-powered aircraft into the Polish air force, all Yak-23s were allocated to the Cracow-based

Polish pilot in Yak-23 cockpit. Details of the canopy and gun-sight are visible.

Author's coll.

Yak-23 of OSL-4 at Radom, Polish Air Force in flight,
Author's coll.

7th Fighter Aviation Division commanded by the excellent and outstanding Col. Jan Frey-Bielecki. After conversion training of all regiments of the division was completed and it reached operational capability, the commander and his staff prepared a tactical exercise to develop methods of group air combat by a regiment. The exercise was going to involve the 2. and 39. PLMs, based at Cracow and at Mierzęcice, respectively. The group air combat was going to be fought over Pustynia Błędowska area. Two flights took off from each airfield, each flight with 8 Yak-23s. The Cracow-based regiment flights were led by Capt. Marian Bondzior, and individual sections by T. Parol, Z. Orczyk, J. Jegrowski and others. The formation included T. Mielnicki, A. Stojowski, T. Bakalarski, W. Jędrych, S. Wdowczyk and S. Żylski. The regiment from Mierzęcice was led by Capt. Z. Ulanowski, with section leaders: B. Kuczyński, Cz. Dłużyński, P. Rojek. Other pilots included J. Majorowski, K. Gielas, Z. Czech, A. Śliwiński, M. Maj, S. Młynek, S. Steczek, F. Komendacki and J. Brzosko.

Approach to the combat area was flown by flights and sections in line astern. Col. J. Frey-Bielecki and Maj. A. Rybacki from the 7. DLM OPL HQ were the exercise referees and supervisors, they flew in a pair of Yak-23s side-by-side with the Cracow formation. For safety individual regiments were assigned different cruising altitudes for their approach to the head-on encounter area, and different parts of the combat area (northern and southern portion of Pustynia Błędowska). In reality leaders of both formations, in order to achieve better attack positions, up sun and with the advantage of height, left the prescribed area by some 20 km, gaining height to 5,000 m. After a few minutes of flying they met head-on. Both groups simultaneously turned to starboard. The line of sections transformed into a circle. At the time there were some clouds in the area.

Pilots changed their altitudes, to avoid falling into clouds. In this situation all formations broke loose. After a few minutes individual aircraft chased each other, attempting to aim accurately and photograph the enemy, whichever regiment he was from. Chaos reigned. In the tumult of radio communications, the voice of the division commander, well known to all pilots, broke through:

"I Thunder – cease fighting. Return to base". Fortunately, the uncontrolled combat was interrupted in time. All crews returned safely to their bases. The next day the exercise was debriefed. There were no victors. Review of camera gun films revealed no effective firing positions. One film was a sensation: it showed the target under attack, and much nearer on one side another aircraft could be seen, as if trying to ram the enemy. Fortunately there was no collision. In the days that followed the regiments prepared for a repeated exercise. Preparations were more accurate and precise, pilots had to practice all elements on foot at the airfield in their pre-planned formations and individually. The next combat of the regiments was ordered and controlled by the regiment commanders.

Apart from its advantages, the Yak-23 had its shortcomings, such that it failed to enter widespread service. It was a transition period aircraft, in which many pilots gained their first experience in flying jet-powered aircraft. Its tactical performance was poor, flight endurance with two additional tanks virtually not exceeding 1 hour. During flights with speeds of Ma 0.8 or above it suffered from directional instability. When opening the throttle the nose of the fuselage went immediately up, making it difficult for the pilot to watch the space ahead. This shortcoming and lack of air brakes prevented effective manoeuvring, and this made air combat significantly more difficult. The tests conducted by the Americans confirmed these features of the aircraft, allowing them to develop proper combat tactics. All this meant that in the late 1940s and early 1950s the design was used in the USSR only as a transition fighter.

Polish pilots who flew Yak-23s learned its features well, and some of these became evident after conversion onto MiG-15s. During Yak-23s flights to extreme heights the cockpit canopy invariably froze over, resulting in limited visibility. Sometimes elements of the control system froze over. Another phenomenon was a swelling stomach, resulting in buttons the flying suit jacket tearing loose. Doctors and scientists did not want to believe that it was possible to fly such missions at temperatures of -56° with just the oxygen mask. Most pilots considered the aircraft to have little combat potential, but to be a good air combat training device. However, rapid development of aircraft engineering

Polish fitter and Yak-23 of 2 PLM, Kraków, after flight.

R. Gretzyngier coll.

in the early 1950s meant that many designs failed to keep up and had to be phased out in favour of better and more modern ones. This happened to the Yak-23, used for just a few years in the USSR, while its younger competitor remained in service ten times longer.

Bibliography:

- Bartoszewski Piotr, "Samolot myśliwski Jak-23", MON, Warsaw 1986
- Gordon Yefim, "Early Soviet Jet Fighters", Midland Publishing, Hinckley 2002
- Kmiecik Tadeusz, "Polskie Lotnictwo Wojskowe 1945-1962", ULMAK, Warsaw 2001
- Kopenhagen Wilfried, "Sowjetische Jagdflugzeuge", Transpress, Berlin 1984
- Kopenhagen Wilfried, "Das Große Flugzeug Typenbuch", Transpres, Berlin 1987
- Królikiewicz Tadeusz, "Polski samolot i barwa", MON, Warsaw 1981
- Marciniuk Andrzej, "Oficerska Szkoła Lotnicza nr 5 im. Franciszka Żwirki i Stanisława Wigury w latach 1951-1964", Polish National Defence Academy, Warsaw 2001
- Mizera Józef, "Skrzydlate lata", Zespół Redakcyjno-Wydawniczy WLOP, Poznań 1999
- Morgała Andrzej, "Polskie samoloty wojskowe 1945-1980", MON, Warsaw 1981
- Popioł Waldemar, "Dzieje 11. Dywizji Lotnictwa Myśliwskiego w latach 1951-1984", Wydawnictwo Politechniki Koszalińskiej, Koszalin 2002
- Krumbach Jan, Vrany Jiri, Hurt Zdenek, "Ilustrovana historie letectvi", Nase Vojsko, Prague 1986
- Zieliński Józef (ed.), "Pamięci lotników wojskowych 1945-2003", Bellona, Warsaw 2003
- Avyatsya i Vremya no. 2/2003, Gordon Yefim, "Yak-23 Istrebityel na eksport", Kiev, 2003
- Avyatsya i Vremya no. 6/2002, Gordon Yefim, "Reaktivniye pervency Yakovlev", Kiev, 2002
- Lotnictwo no. 1-2/2003, Bartosik Sławomir, Łaz Marek, Mikołajczuk Marian, Senkowski Robert, "Pierwsze samoloty odrzutowe w lotnictwie polskim" pt. I, II, Warsaw 2003
- Lotnictwo no. 12/2004, Gruszczyński Jerzy, Fiszer Michał, "Encyklopedia lotnicza", Warsaw 2004
- Lotnictwo z szachownicą no. 6/2002, Andrzej Tatarek, "Odrzutowa Agata czyli Jakowlew Jak-17W", Wrocław 2002
- HPM no. 9/2001, Ina Miroslav, "Jakovlev Jak-23 (S-101) u Cs. Letectva", Prague 2001
- HPM no. 7/2003, "Jakovlev Jak-23 (S-101) u Cs. Letectva, post scriptum", Prague 2003
- HPM no. 9/2001, Khol Miroslav," Letadla nasich muzei – Zahada HX-51", Prague 2001
- Letectvi+Kosmonautika no. 9/1984, Balous Miroslav, "Jakovlev Jak-15/21", Prague 1984
- Armady Technika Militaria no. 4/2006, "Boj o letecke technologie za studene valky" (4), Prague 2006
- Krilya Rodiny no. 4/1998, Vasilyev Nikolai, Podolniy Yevgeniy, "Na starte reaktivnogo marafona", Moscow 1998
- Internet, Makedonska Awiacja Forum, Ukradeniut Yakovlev 23 Flora, Cedomir Janic
- Internet, Air Seria Forum

Detail photos

Yak-15

Yak-15, ¾ front view.

Port front fuselage of Yak-15, showing cannon barrel and engine cowling locks.

Both Dmitri Sribnyi.

Yak-15, starboard front fuselage.

Engine air intake, and both cannon barrels.

Close-up of engine air intake, showing compressor blades and intake cone.
All Dmitri Sribnyi.

Starboard centre fuselage, note the wing-fuselage fairing.

Port centre fuselage.

Engine nozzle covered with temporary wooden cover.
All Dmitri Sribnyi.

Top: Cockpit canopy, left view; note gunsight.

Three photos showing details of cockpit.
All Dmitri Sribnyi.

Left: *Close-up of gunsight, front view.*

Below: *Close-up of cannon barrel outlet fairing.*

Left: *Front view, showing shape of upper part of fuselage.*

Bottom: *Two photos showing canopy and navigation system cover details* ***All Dmitri Sribnyi.***

Antenna mast fastening, starboard fuselage side.

Lower part of engine cowling, starboard side.

Lower starboard part of fuselage at the wing leading edge.
All Dmitri Sribnyi.

Two shots of port fuselage centre.

All Dmitri Sribnyi.

Starboard wing, view from above, showing inboard part of aileron.

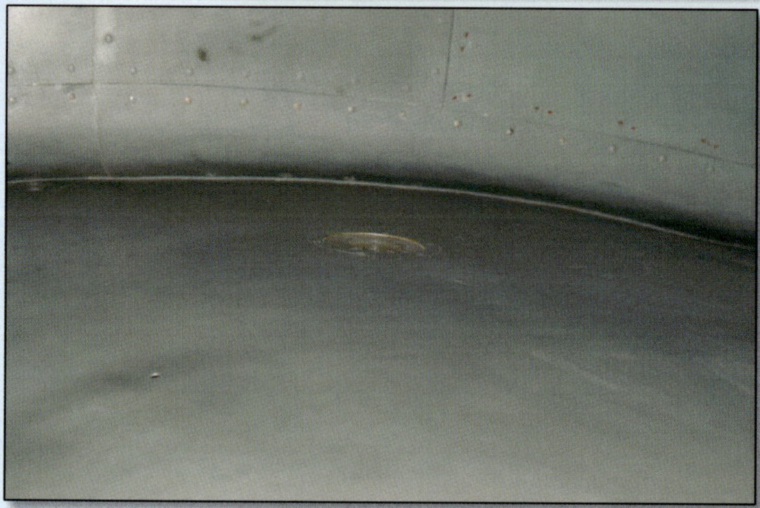

Port wing root; note fuel gauge shown in detail photo at right.
All Dmitri Sribnyi.

Starboard aileron showing ground-adjustable trim tab.

Starboard wing root fairing.

Starboard wingtip and navigation light.
All Dmitri Sribnyi.

Details of vertical tail-plane.
All Dmitri Sribnyi.

Starboard (top) and port (bottom) tailplane. Rudder actuator (centre).

All Dmitri Sribnyi.

Main undercarriage wheel, view from inboard.

Port main undercarriage leg, front view.
All Dmitri Sribnyi.

Above: *Two shots of starboard main undercarriage leg.*
Bottom: *Port main undercarriage bay.* ***All Dmitri Sribnyi.***

Main undercarriage jack.

Details of main undercarriage bay

Tailwheel. Note deflector protecting the wheel against exhaust gas.
All Dmitri Sribnyi.

Yak-17

Yak-17, left front view. This a/c preserved in the Letecké Muzeum Kbely. Serial number IS-10001.
Right front view of Yak-17.

All photos Stratus coll.

Yak-17, ¾ front view.

Front fuselage. Engine intake and cannon barrel are clearly visible.

Compressor bleed openings in the lower part of fuselage.

All photos Stratus coll.

Right side of canopy.

Left rear view of tail unit.

Engine air inlet, showing vanes and compressor blades.

All Stratus coll.

Port main undercarriage leg and undercarriage bay, front view.

Port main undercarriage leg, view from outboard.

Port undercarriage bay, showing inboard part of wheel bay door.

All photos Stratus coll.

Left: Starboard undercarriage leg, view from inboard.

Top, right: *Inboard door of main undercarriage bay.*

Bottom: *Port main undercarriage bay.*
All photos Stratus coll.

Top, left: Starboard main undercarriage bay, showing inboard door jack.

Top, right: Starboard main undercarriage leg, front view.

Right: Main undercarriage leg outer door.

Bottom: Two photos of main wheel.

All photos Stratus coll.

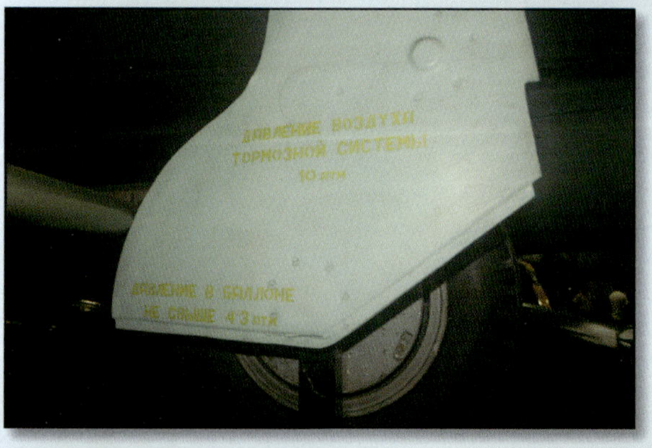

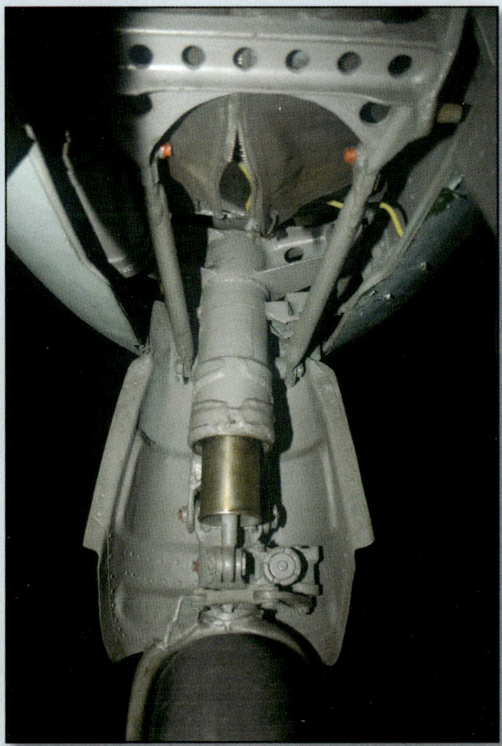

***Top left**: Nose undercarriage, rear view.*

***Top, right**: Nose undercarriage door, front view.*

***Left**: Nose undercarriage, right view.*

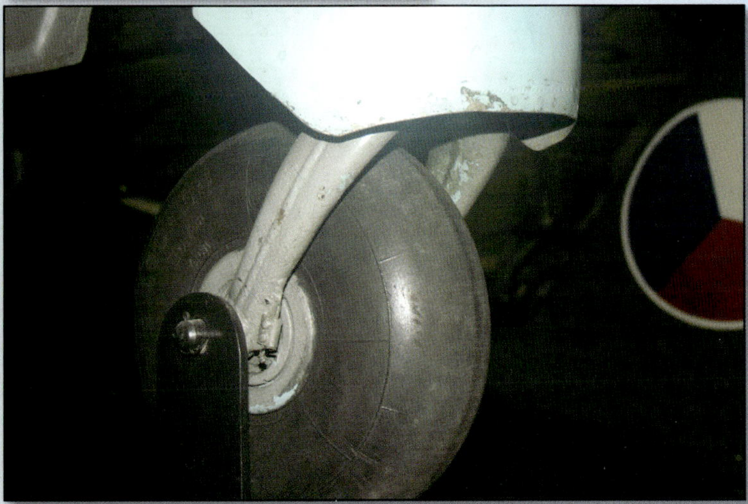

Nosewheel.

All photos Stratus coll.

Yak-17UTI

Top: *Yak-17UTI, ¾ front right view.*

Starboard fuselage centre, rear view.

Fuselage centre and cockpit canopy.
All photos Stratus coll.

Top, left: *Front fuselage and nose undercarriage, right view.*

Top, right: *Air inlet and nose undercarriage doors, front view.*

Left: *Front port part of fuselage.*

Front part of canopy.
All photos Stratus coll.

Port main undercarriage leg, outboard door.

Port main undercarriage leg, front view.

Port main undercarriage leg, inboard view.
All photos Stratus coll.

Tailplane unit, right view.

Starboard tailplane.

Starboard fuselage centre below the canopy.
All photos Stratus coll.

Yak-23

Top: *Yak-23, the a/c preserved as a monument in Zakopane, right view.*

Left: *Front port part of fuselage.*

Bottom: *Yak-23, left view.*

All photos Stratus coll.

Yak-23, front view. ***Stratus coll.***

Subassemblies of the Yak-23. Drawing from Technical Manual

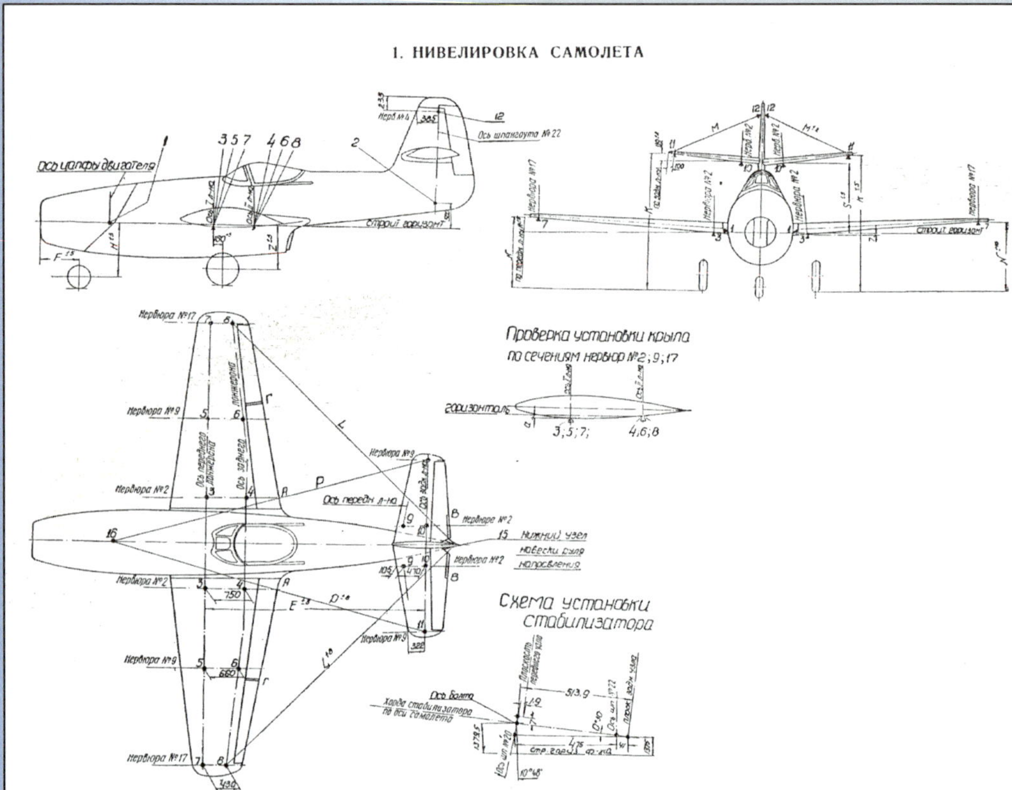

Yak-23, levelling plan. Technical Manual

Yak-23 of the Polish Air Force, ¾ front view. **A. Gołąbek.**

Yak-23, view from above. The a/c preserved as a monument at one of housing estates in Stalowa Wola, Poland, in the 1960s.

Muzeum Regionalne Stalowa Wola via A. Garbarz

Left: *Port fuselage centre.*

Bottom: *Left view of the Yak-23 preserved in the MLP Cracow. The a/c is fitted with wingtip fuel tanks.*

All photos Stratus coll.

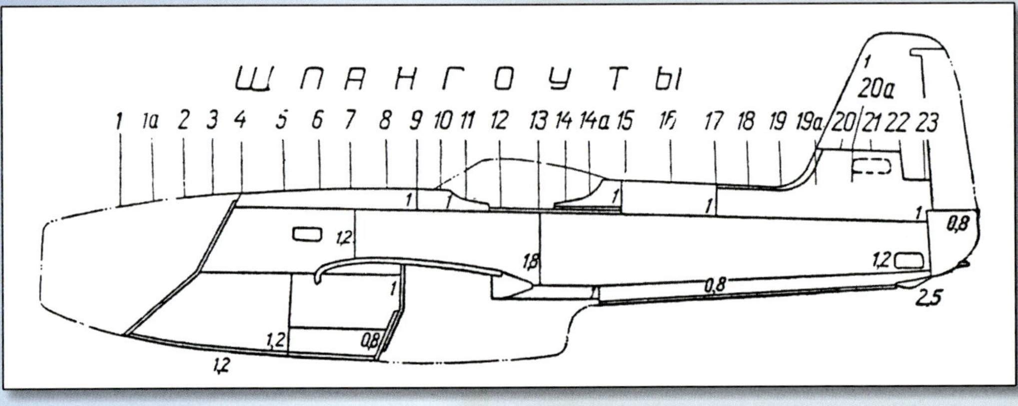
ШПАНГОУТЫ
1 1a 2 3 4 5 6 7 8 9 10 11 12 13 14 14a 15 16 17 18 19 19a 20 21 22 23
20a
1,2
1,8
0,8
2,5

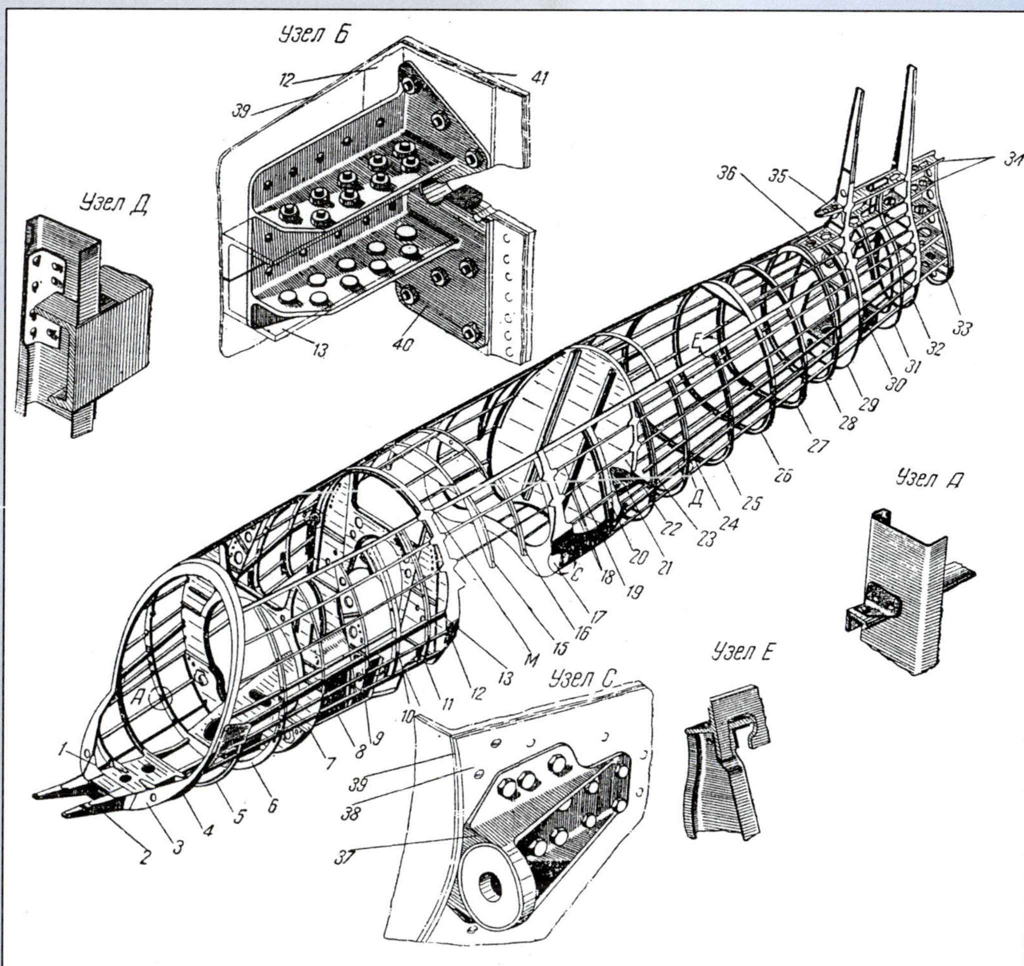

Opposite page: *Yak-23, front view (top) and rear view (centre). Stratus coll.*
Opposite page, bottom: *System of fuselage frames and their numbers.*

This page:
Fuselage structure and attachment points.
Technical Manual

Left: *Front starboard fuselage.*
Stratus coll.

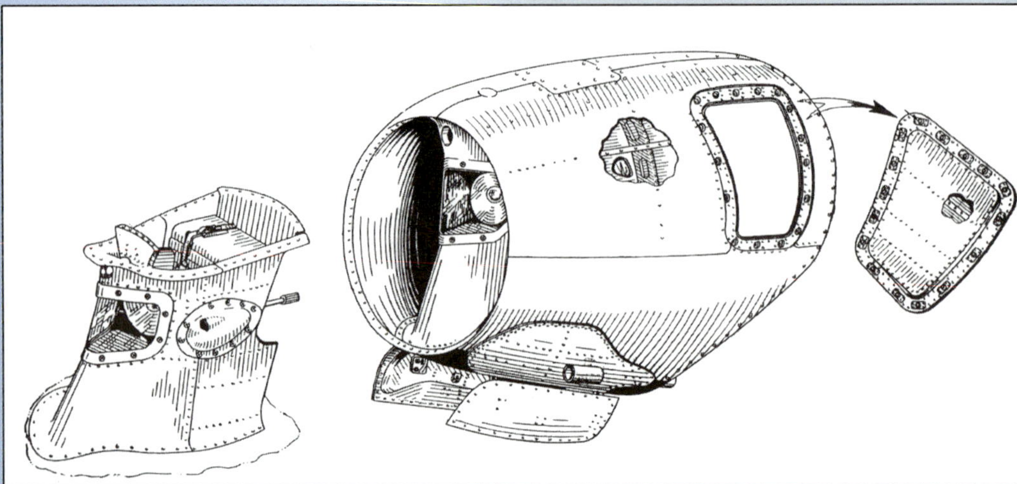

Top: *Front removable engine cowling and details of landing light.*
Technical Manual

Right: *Engine air intake and landing light.*
Stratus coll.

Front fuselage, port side. The removable engine access panel can be seen.
Stratus coll.

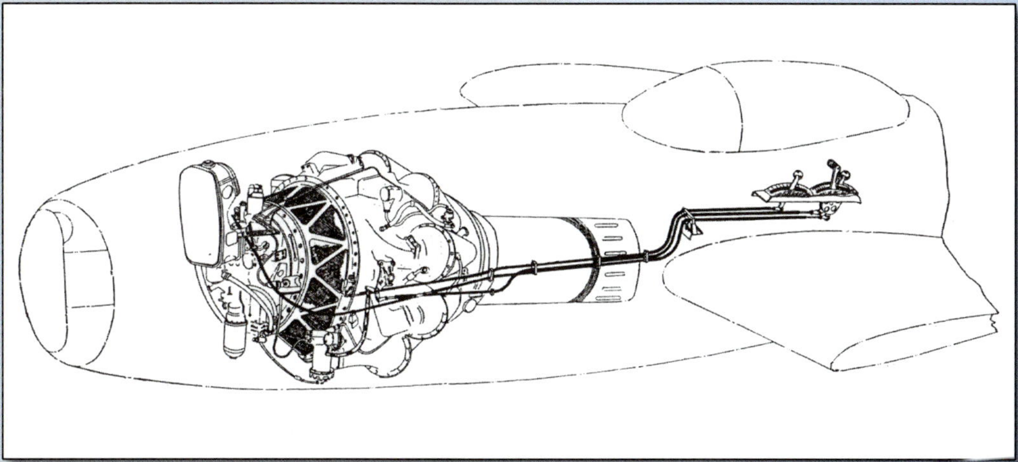

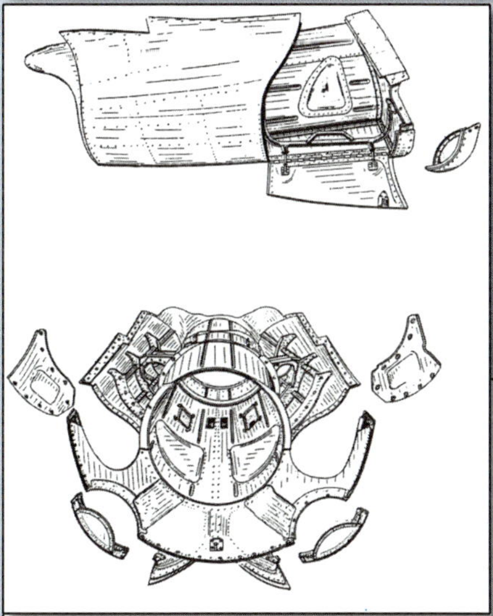

Top: *Engine installation in the fuselage.*

Left: *Rear engine cowlings.*

Technical Manual

Bottom: *Yak-23 of Polish Air Force at Tomaszów Mazowiecki airfield with nose removed for engine maintenance, summer 1961.*

F. Dziekan coll.

3 shots of engine intake and details at the port side of air intake divider.

Engine nozzle and details of wing root fairing.

All photos Stratus coll.

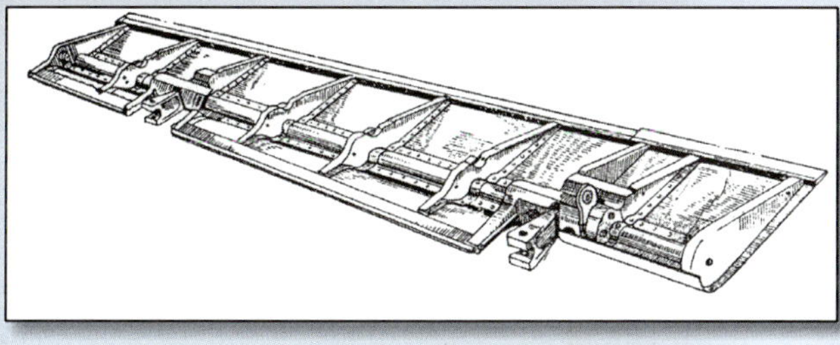

Two photos of upper surface of wings.

Stratus coll.

Left: *Aileron structure.*

Bottom: *Port wing structure and removable wingtip.*

Technical Manual

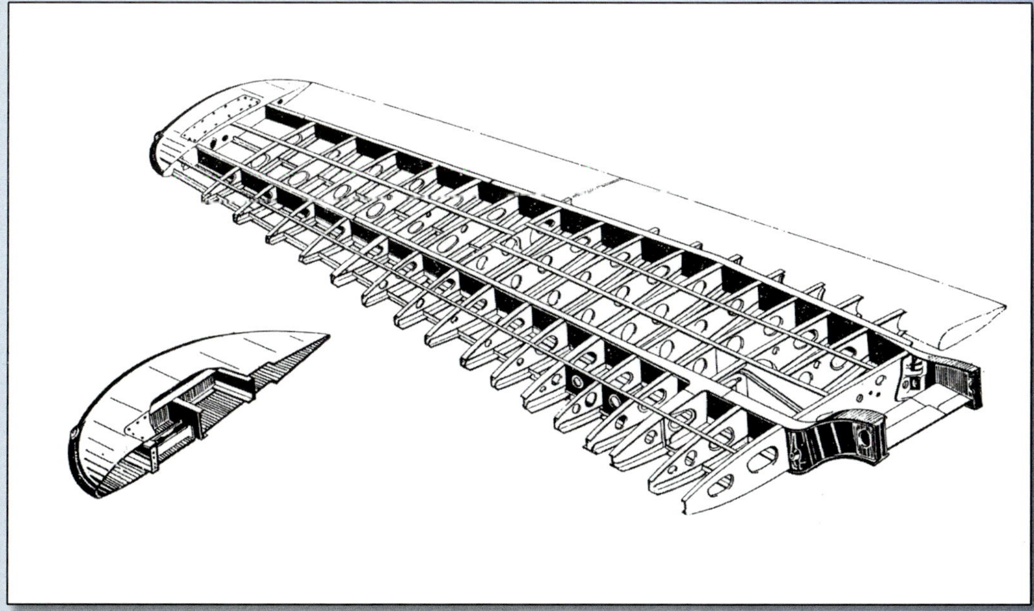

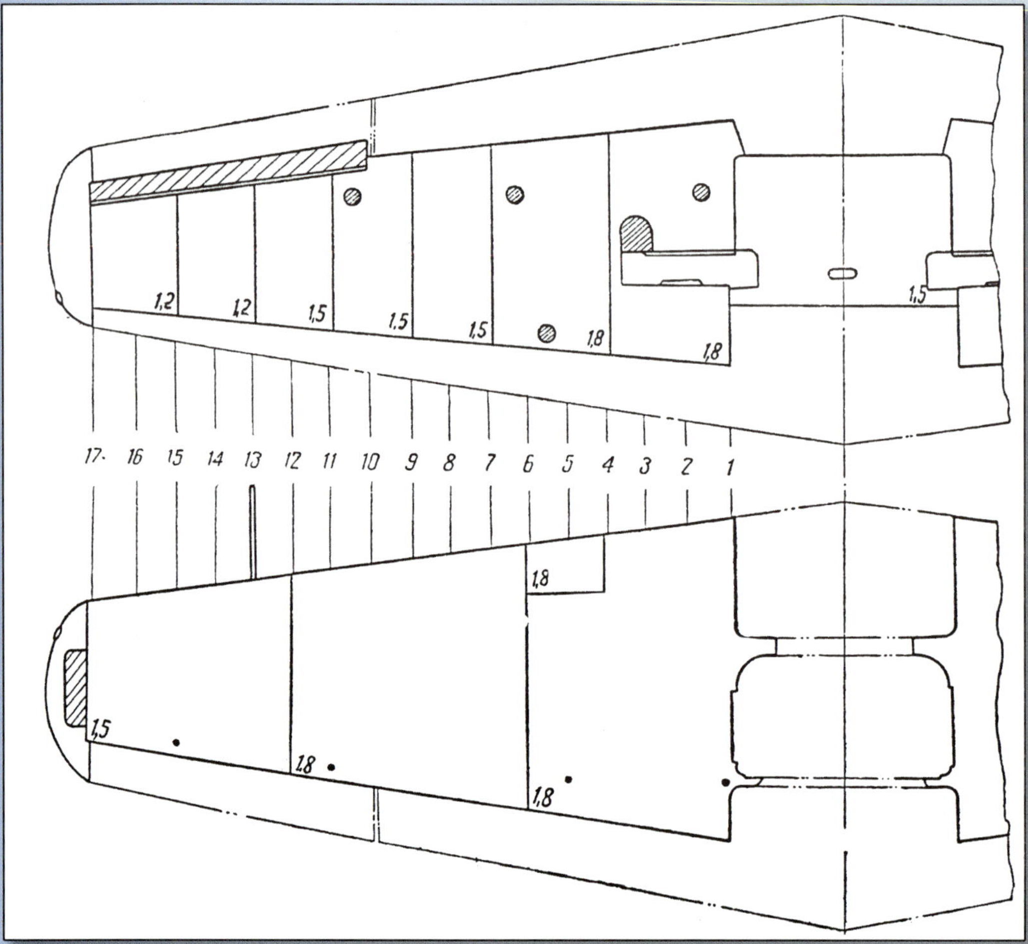

Arrangement of wing ribs and wing panels.

Technical Manual

Starboard wing root. Note the wing walkway.
Stratus coll.

Port wing leading edge at the fuselage. Note the walkway.

Starboard wing root fairing.

Pitot probe in port wing.
All photos Stratus coll.

Top, left: Tail unit, front view.

Top, right: Tail unit, right view.

Right: Tailplane structure.
Technical Manual

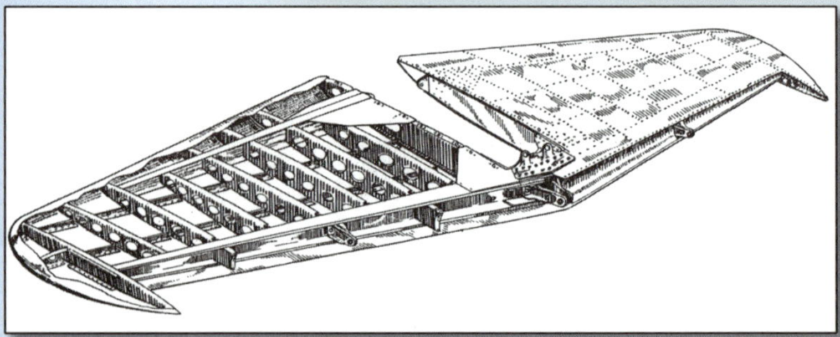

Bottom: Rear starboard part of the a/c.
All photos Stratus coll.

Tail unit details of the Yak-23.

All photos Stratus coll.

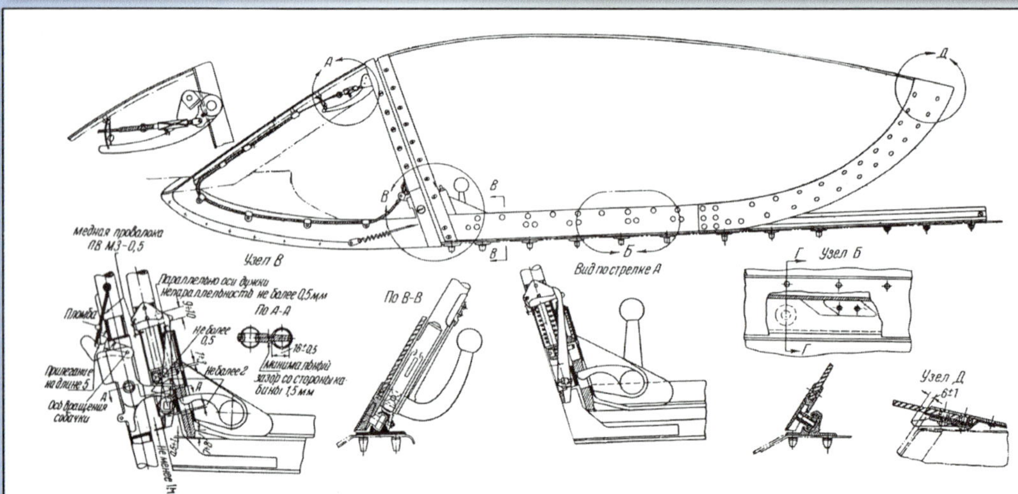

Top: *Port side of canopy. Note armoured glass frame.*

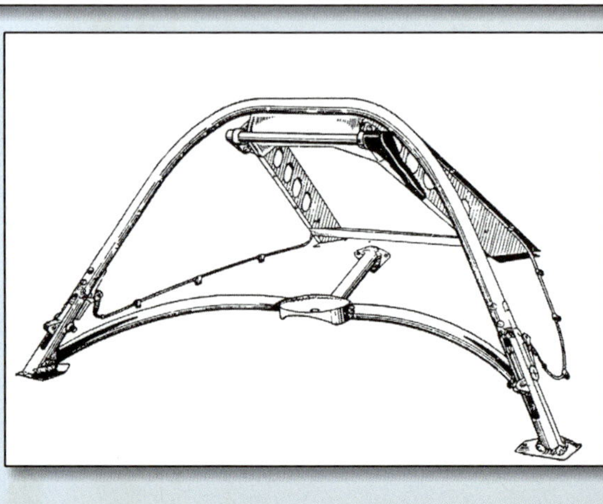

Drawings showing details of cockpit canopy.
Technical Manual

Left: Cockpit canopy, front view.

Middle: Canopy, left view.

Middle, right: Canopy in open position, right view.

Bottom: Windscreeen, front right view.

All photos Stratus coll.

Instrument panel. Note also gunsight and control column.
Mark I Publication coll.

Above: An original photo showing instrument panel and port cockpit panel.
Author's coll.

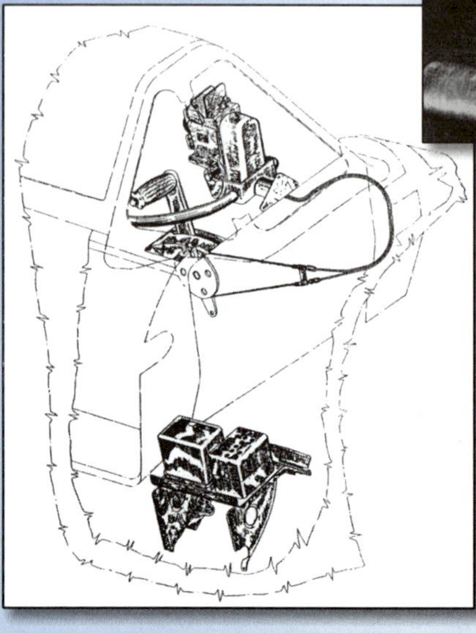

Drawing of the gunsight.
Technical Manual

Starboard cockpit wall.

Port cockpit wall.

Details of pilot seat and control column. Note also the rudder bar.
All Mark I Publication coll.

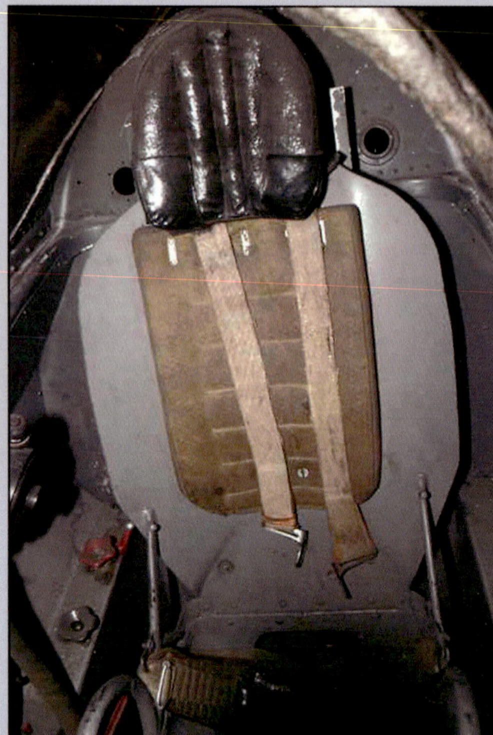

Close-up of the gunsight. ***Mark I Pub.***

Pilot seat, showing headrest and belts. Mark I Pub.

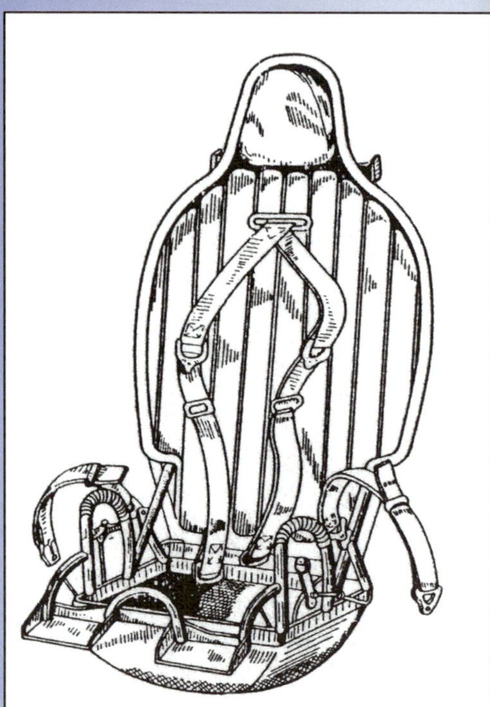

Two drawings showing pilot seat, front and rear view. Technical Manual

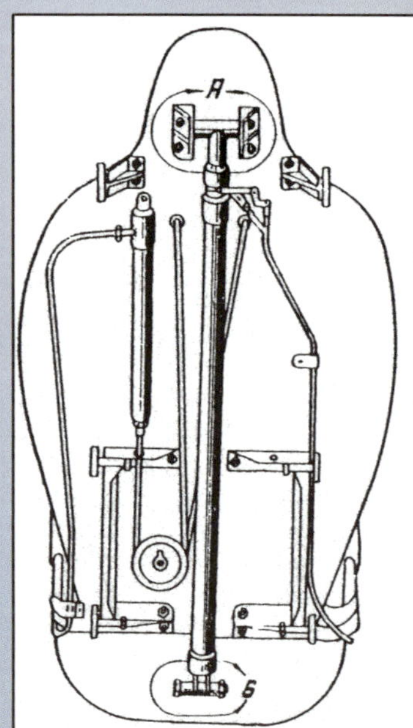

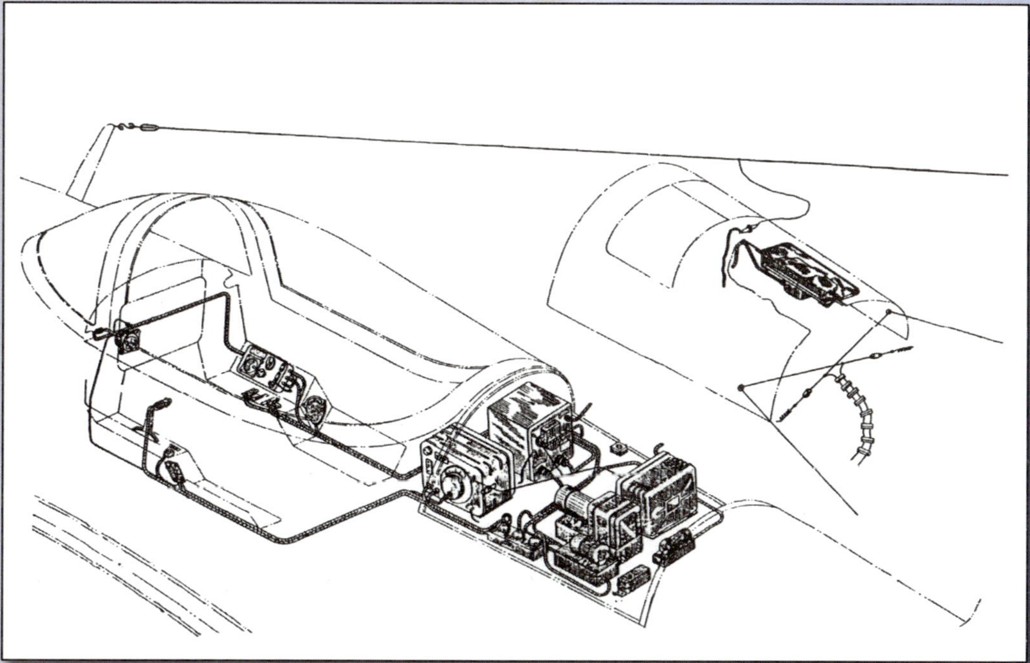

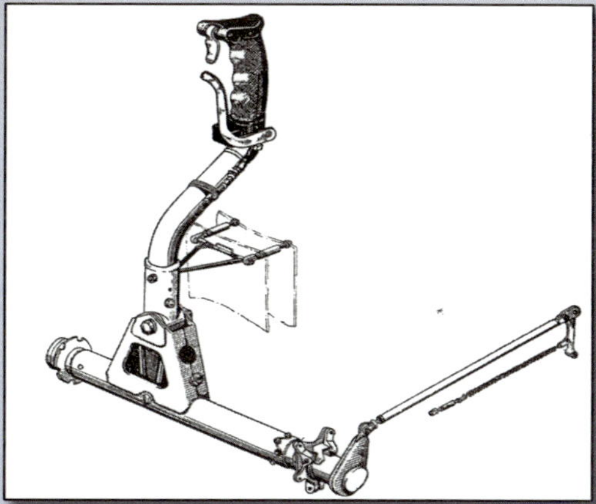

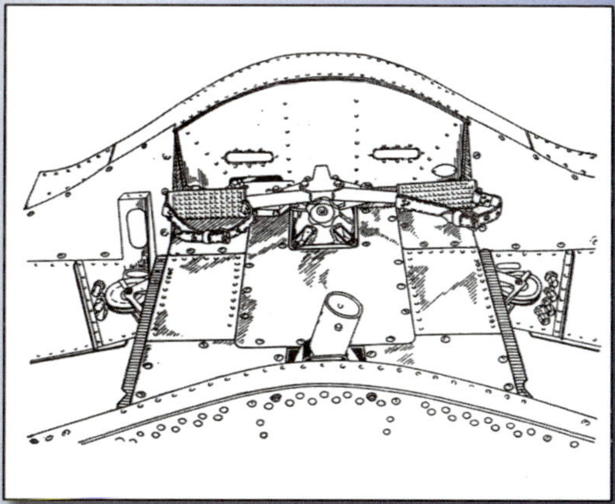

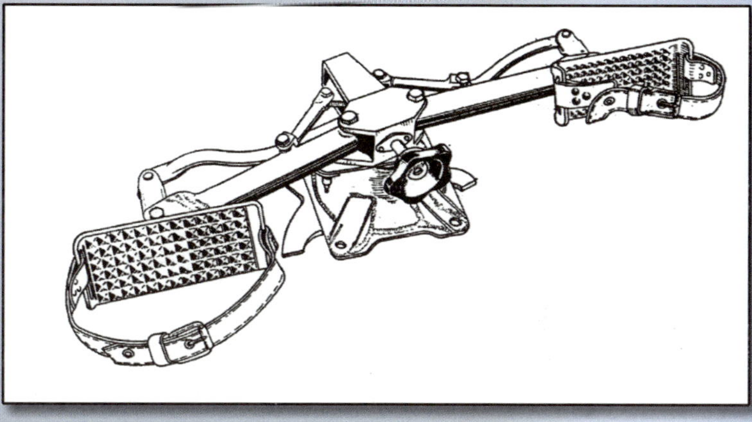

Top: *Radio equipment.*

Middle, left: *Details of control column.*

Middle, right: *Cockpit floor and rudder bar attachment.*

Bottom: *Details of the rudder bar.*

Technical Manual

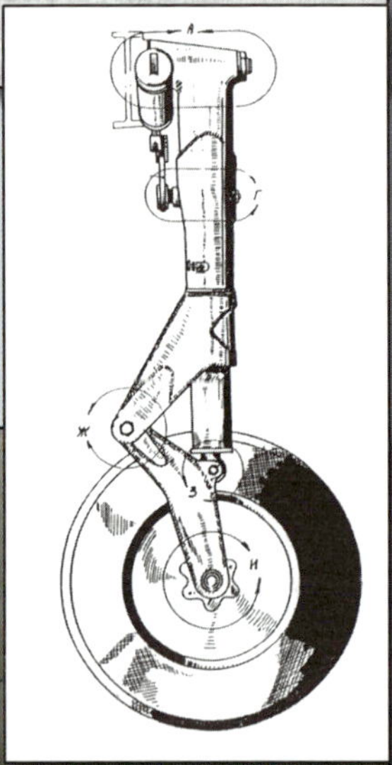

Top: Yak-23, right front view. The a/c fitted with wingtip fuel tanks. Note the undercarriage arrangement. ***Author's coll.***
Right: Main undercarriage leg. Technical Manual
Bottom: Starboard main undercarriage leg, 2/3 front view. ***Stratus coll.***

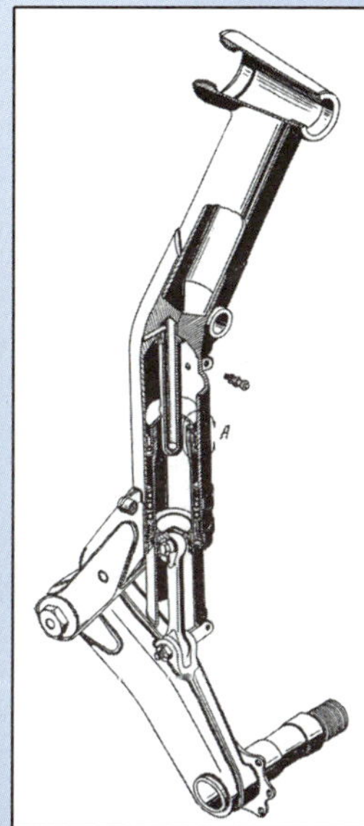

Left: Starboard main undercarriage leg, front view. Stratus coll.
Top: *Details of main undercarriage leg. Technical Manual*

Bottom, left: *Port undercarriage leg doors*
Stratus coll.

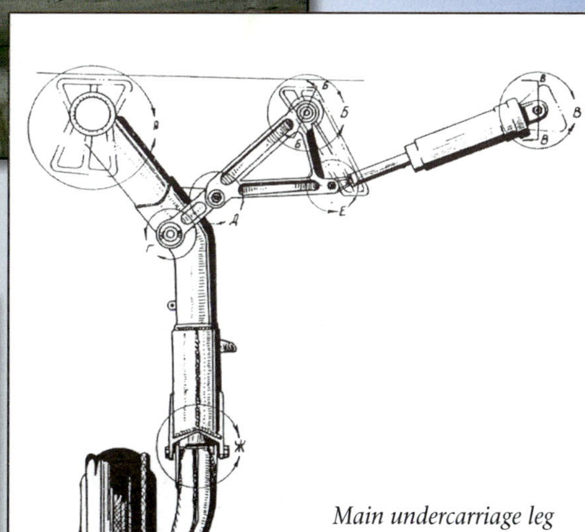

Main undercarriage leg jack.
Technical Manual

Port main undercarriage bay.

Main undercarriage inboard doors.

Two inboard views of the main undercarriage.
All photos Stratus coll.

Top: Starboard undercarriage leg, rear view. Shown also is the engine nozzle.
Stratus coll.

Details of nose undercarriage.
Technical Manual

Nose undercarriage, right view.

Nose undercarriage, left view. Note u/c doors and cannon barrel.

Nose wheel.
All photos Stratus coll.

Yak-15 with Jumo engine, the first prototype, late version. Moscow, December 1945.

Yak-15RD, the second prototype during manufacturer's test flight, pilot M.I. Ivanov, Moscow, April-June 1946.

Yak-15, 31002, (yellow 32 T) during State trials, pilots G.A. Sedov and A.G. Proschakov, April 1947.

Yak-15 (yellow 37), unarmed aircraft of the "parade" group, Maj. Caucausus Ridge, Moscow, 1st May 1947.

Yak-15 (yellow 2) an aircraft of the aerobatic group, Tushino, 1947.

Yak-15 (yellow 20 T) used in the Soviet Air Forces as a training aircraft, late '40s.

Yak-15 (red 2) of Lt. Dimitri Samoylov, 139th 'Smolensk' Guards Fighter Regiment, 303th Fighter Divisjion Kobryn, 1948.

Yak-15 (yellow 47) used as a testbed for wing-to-wing in-flight refuelling system, summer 1949.

Yak-15 (yellow 58) of 2nd Squadron, 3rd Guard Fighter Regiment, Cherlany, 1948.

Yak-15 (yellow 89 1), test bed aircraft, 1946.

Yak-15 (yellow 57 2) during checkout trials at the Red Banner State research Institute of the Soviet Air Force, 1947.

Yak-15 (yellow 12 2) an early production aircraft at No. 31 Factory, Tibilisi, 1947.

Yak-15 (yellow 37), aircraft preserved on display at the OKB Yakovlev Museum, Moscov, Chodynka, 1989.

Yak-17 (white 45) Soviet Air Force, late '40s.

Yal-17 (Black 02) Soviet Air Force, aircraft in markings as preserved in Central Russian Air Museum in Monino.

Yak-17 (red 27) of Soviet Air Force, late '40s.

Yak-17 (yellow 20) of Soviet Air Force, late '40s.

Yak-17, S-100, (yellow 30) as seen at Mlada Bolesav factory. The only Yak-17 in Czechoslovak markings.

Yak-17, (yellow 29) of 1st Fighter Aviation Regiment, Polish Air Force, early '50s.

Yak-21 prototype (yellow 101), May 1947.

Yak-17UTI, (yellow 07) Soviet Air Force, early '50s.

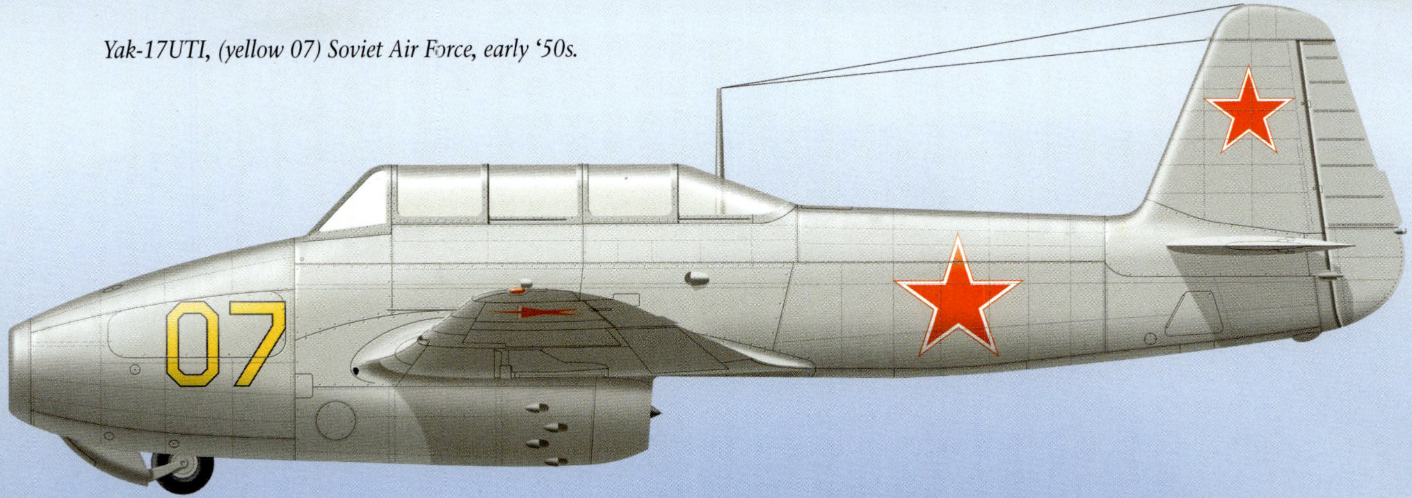

Yak-17UTI, (yellow 19) Soviet Air Force, early '50s.

Yak-17UTI, (red 41) Soviet Air Force, early '50s.

Yak-17UTI (red 12) of PLAAF, early '50s.

Yak-17UTI (blue 60) of 29 GIAP, Chyanan, PLAAF, autumn 1950.

Yak-17UTI (red 02) of Polish Air Force, early '50s.

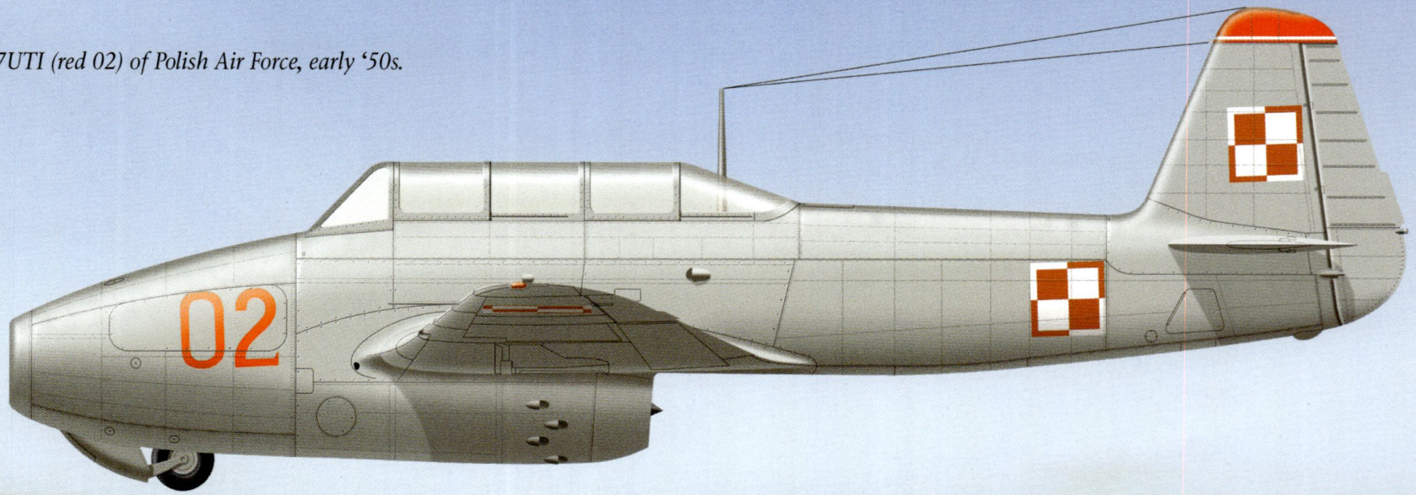

Yak-17 W, SP-GLM one of the few civil-registered Yak-17s used for research purposes by the Instytut Lotnictwa (Polish Aviation Institute).

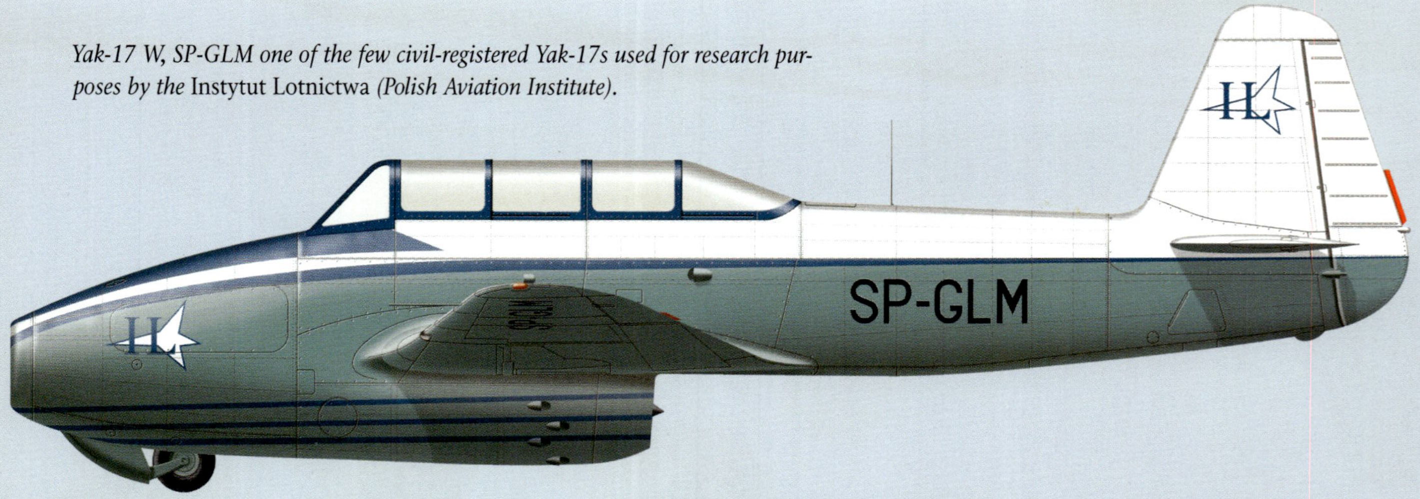

Yak-17 W (yellow 1), serial 3120132 at Radom, 1953.

Yak-17 W (blue 7) of Romanian Air Force, early '50s.

Yak-23 (red 118) of Bulgarian Air Force, mid '50s.

Yak-23 (yellow 23) of Bulgarian Air Force, mid '50s.

Upper and Lover views of
Bulgarian Yak-23.

Yakovlev Yak-23

Two Yak-23s (red 41 & red 12) of Bulgarian Air Force, mid '50s.

Yak-23 (yellow 68) of Bulgarian Air Force, mid '50s.

Yak-23 (red 14) of Bulgarian Air Force. Late '50s.

Yakovlev Yak-23

Upper and lower views of Czechoslovak Yak-23.

Yak-23, S-101, *Flora*, IF-11, of 1st Squadron of the 51st Air Regiment based at Milovice-Mladá. Mid '50s.

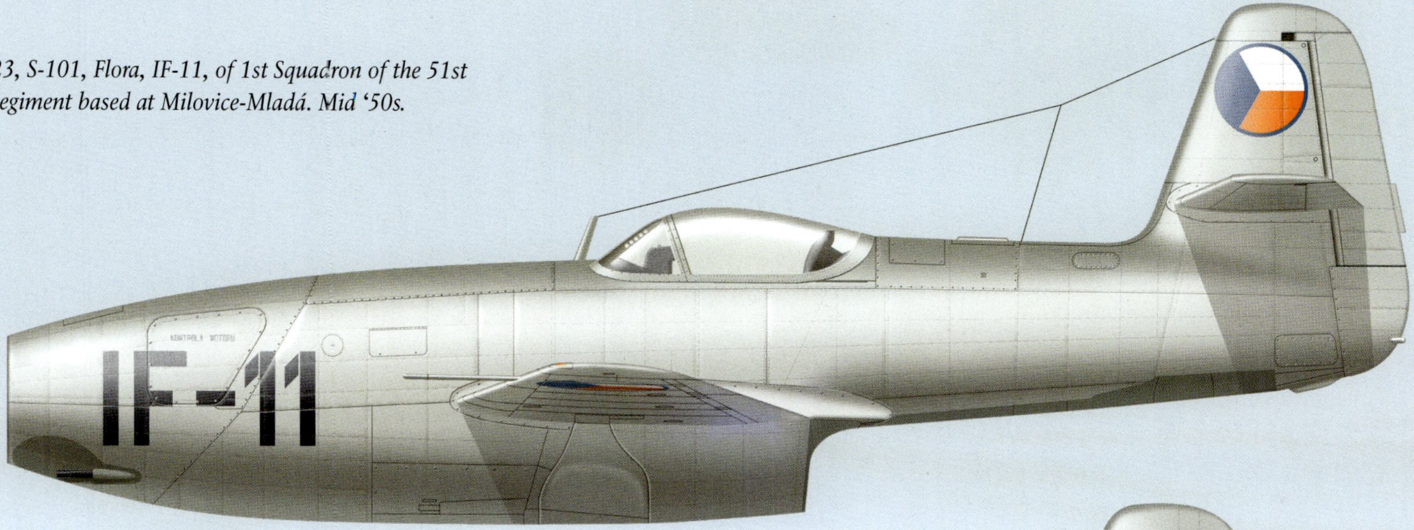

Yak-23, S-101, *Flora*, KR-17 of VSL Air Force Jet Training Centre at Milovice-Mladá, mid '50s.

Yak-23, S-101 Flora, HX-51 (s/n 10101) of 3rd Squadron, 3rd Air Regiment. mid '50s

Yak-23 (red 04) of Polish Air Force, Radom airfield, 1956.

This and opposite page: Yak-23 (red 09) of Polish Air Force, mid '50s.

Yak-23 (red 11) of Polish Air Force, mid '50s.

Yak-23 (red 24) of Polish Air Force, mid '50s.

Yak-23 (blue 121) of OSL-5 Radom. Polish Air Force , mid '50s.

Yak-23 (blue 144) of OSL-5 Radom. Polish Air Force, mid '50s.

Yak-23 (SP-GLK), the only civil Yak-23 in the world, used by Instytut Lotnictwa in Warsaw (Polish Aviation Institute), late '50s. Andrzej Abłamowicz set two time-to-height records in this aircraft on 21st September 1957.

Yak-23 (yellow 07) of OSL-5 Radom. Polish Air Force , mid '50s.

Yak-23 (yellow 11) at Tomaszów Mazowiecki airfield. Polish Air Force, mid '50s.

Yak-23 (yellow 12) of OSL-5 Radom. Polish Air Force, mid '50s.

Yak-23 (yellow 16) of OSL-5 Radom. Polish Air Force, mid '50s.

Yak-23 (red 010) of Polish Air Force, mid '50s.

Yak-23 (red 23) of Romanian Air Force, late '50s.

Yak-23 (red 28) of Romanian Air Force, late '50s.

Yak-23 during tests in the USA with American markings and "FU-599" code.

Yak-23 (yellow 01) during trials at GK NII WS (Soviet Air Force).

Yak 23 (yellow 52) of Soviet Air Force, early '50s.